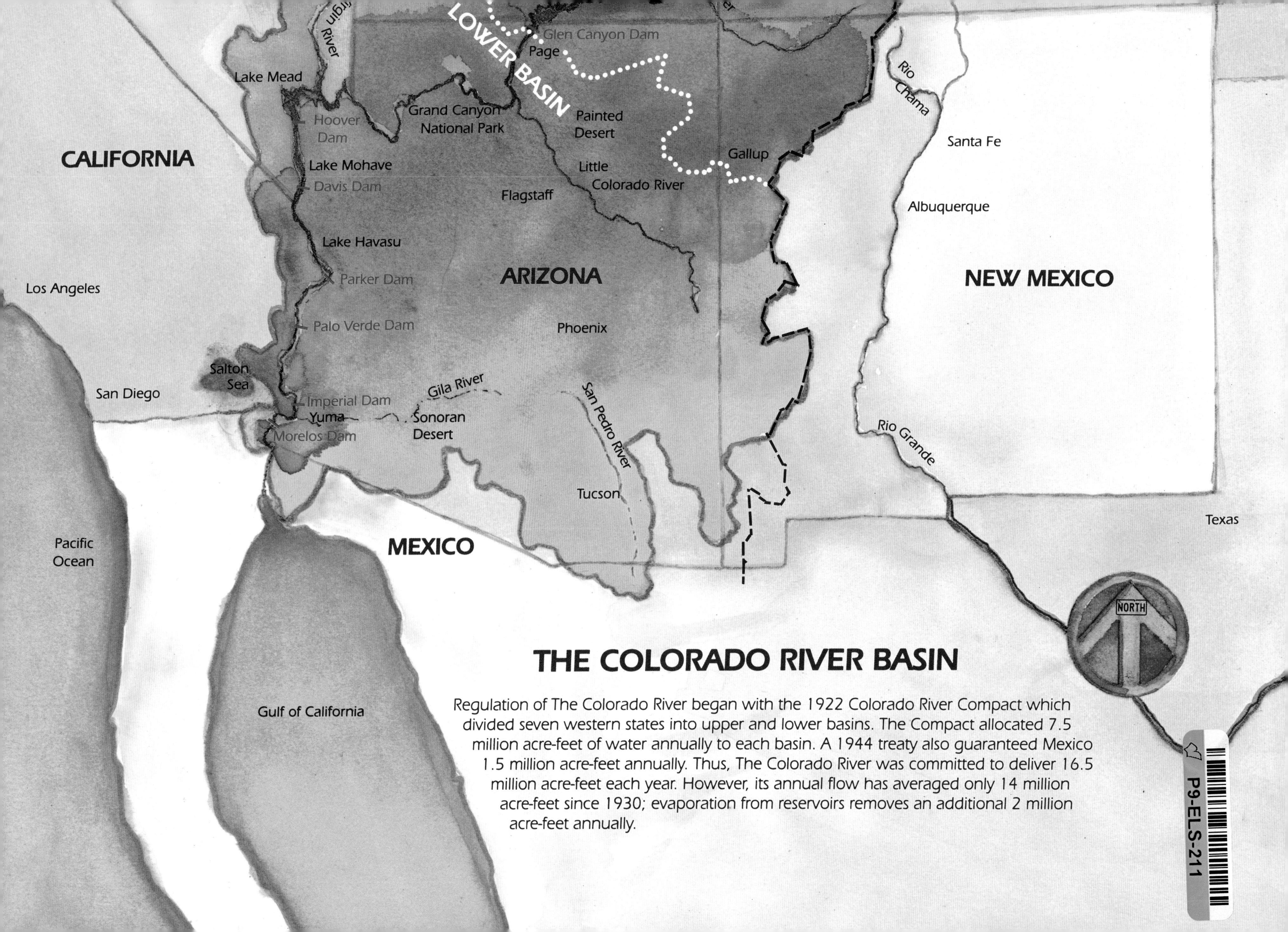

THE COLORADO RIVER BASIN

Regulation of The Colorado River began with the 1922 Colorado River Compact which divided seven western states into upper and lower basins. The Compact allocated 7.5 million acre-feet of water annually to each basin. A 1944 treaty also guaranteed Mexico 1.5 million acre-feet annually. Thus, The Colorado River was committed to deliver 16.5 million acre-feet each year. However, its annual flow has averaged only 14 million acre-feet since 1930; evaporation from reservoirs removes an additional 2 million acre-feet annually.

THE COLORADO
A River At Risk

Produced by Rich Clarkson

Westcliffe Publishers, Inc.
Englewood, Colorado

Here, in the Never Summer Range, the first water of the Colorado River gathers in the gentle pools of a mountain meadow.

A fantasyland in the desert, Lake Powell's thousands of water-filled canyons attract more than three million visitors a year

Labor Day party boats fill the popular, but tiny, Copper Canyon on Lake Havasu.

Sparkling in the growing dusk, Hoover Dam is the jewel in the Bureau of Reclamation's Colorado River system.

This money-green, irrigated field in the Imperial Irrigation District produces vegetables all year round.

THE COLORADO

A RIVER AT RISK

The headwaters of the Colorado River in Rocky Mountain National Park.

PHOTOGRAPHS BY JIM RICHARDSON

TEXT BY JIM CARRIER

Published by WESTCLIFFE PUBLISHERS, INC.

Editors: RICH CLARKSON, JOHN FIELDER
Designer: KATE GLASSNER
Copy Editors: JOE BULLARD, ANTHONY CAMPBELL
Production Manager: MARY JO LAWRENCE

Foreword adapted by permission and courtesy of Wallace Stegner.
The full piece, "A Geography of Hope,"
may be found in *A Society to Match the Scenery*.
Niwot, Colorado: University Press of Colorado, 1991.

2650 South Zuni Street, Englewood, Colorado 80110
Printed in Singapore by Tien Wah Press (Pte.), Ltd.

International Standard Book Number:
ISBN 0-929969-09-X
LC# 91068438

Contents

FOREWORD

Aridity has been a difficult fact for Americans to accept, and an even more difficult one to adapt to. For nearly the first half of the nineteenth century we avoided the dry country that Pike and Long had called the Great American Desert; but by the 1840s and 1850s we were pushing our agriculture onto the dry plains and trying to mythologize aridity out of existence by such hopeful delusions as that rain follows the plow, that settlement improves the climate. When that myth brought on much human misery and failure as well as much environmental damage, we tried to engineer aridity out of existence by damming and redirecting the rivers. (Most of the results of that plumbing job are still to come, but they are coming.) Or, as a plumbing variation, we robbed Peter to pay Paul: we stole the Owen's Valley's water to make the subdivision of the San Fernando Valley richly profitable; we conducted water through the Rockies from the Western Slope to permit the urban sprawl of Denver.

Human ingenuity has been manipulating Western water for nearly a century, but all our ingenuity has not increased the amount of water or solved any of the secondary problems that lack of water creates. In 1878, a hundred and twelve years ago, John Wesley Powell in his *Report on the Lands of the Arid Region* warned that there was water enough in the West to supply only about a fifth of the land. Being a man of this time, though considerably ahead of it in many ways, he was thinking in agricultural terms, and the new survey and homestead laws that he proposed would have eased the difficulty of Western settlement. Congress, dominated by boosters and local patriots, ignored his proposals, and settlement went ahead by tradition, habit, mythodogy and greed instead of by observation and forethought.

Not even yet has it sunk into some heads that the supply of water is finite. We have water from the time it falls as rain or snow until it has flowed past us, above ground or below, to its final ocean or underground reservoir. We can hold it back and redirect it, but we cannot add one drop to its total supply. In fact, the evaporation rate in really dry country being as much as ten feet a year, we may lose almost as much of it by ponding it as we save by slowing it down.

Moreover, in the West, "using" generally means "using up." What we put to municipal or industrial use is not coming back into the rivers to be available for irrigation – or if it does, it comes back poisoned. What is used in irrigation largely evaporates, and any percentage that finds its way back into the rivers is increasingly laden with salts, fertilizers, and pesticides from the fields. And no matter what use you find for the water from a river, every acre-foot you take out leaves a lessened instream flow to sustain trout, salmon, herons, ducks, rafters, picnickers, skinny-dippers, or whoever else might find it useful. In Colorado, as in some other Western states, you can literally dry up a stream if you have prior rights for some so-called "beneficial" purpose.

Aridity means more than inadequate rainfall. It means inadequate streams, lakes, and springs. It means underground water that replenishes itself more slowly than elsewhere. In some places the underground water is fossil water, irreplaceable within any human time frame. And surface and subsurface water are not two problems, but one.

Confronted by the facts of chronic shortage, the decreasing feasibility of more dams, and the oversubscription of rivers such as the Colorado, the boosters sometimes speak of "augmentation" and suggest ever-greater engineering projects, roughly comparable to the canals of Mars to bring water down to the dry country from the Columbia or the Yukon, or tow it down as icebergs from Glacier Bay to let San Diegans flush their driveways and keep their lawns verdant.

Pipe dreams, arrogant pipe dreams. Why should we expect a desert to blossom? It has, or had until we began to tinker with it, its own intricately interdependent plant and animal species, including the creosote-ring clones that are the oldest living things on earth. The idea of making deserts blossom is something we inherited from Isaiah. It is an idea especially dear to the Mormons, and it has had remarkable short-term success. But it is open to all sorts of questions when we look very far into the future. For financial and political reasons, more than for technical ones, there are limits to how freely we can move water from one watershed to another. There are ecological and (I would say) moral reasons why we shouldn't. As a Crow Indian friend of mine said about the coal in his country, "God put it there: that's a good place for it." Lots of

The Colorado River begins its long journey to the sea in the crystal clear mountain ponds where beaver leave their marks on the tree trunks.

things have learned to depend on the West's water in the places where it naturally occurs. It would become us to leave them their living space, because if we don't, we are taking chances with our own.

Sooner or later we must accept the limitations imposed by aridity, one of the principles of which is a restricted human population.Western growth has a lower ceiling than the boosters are willing to admit, and people in general, both those native to the West and those lured to it by hope or advertising, have so far not welcomed any limitation any more than the boosters have. From very early on, the West has been a land of Cockaigne, an Indian Valley line where every day is payday, a Big Rock Candy Mountain where the bluebird sings to the lemonade springs and the little streams of alcohol come trickling down the rocks. Ordinary people, making it by guess and by God, or not quite making it, have always understood Western hardship, but they have been at least as susceptible to dreams as the ambitious and the greedy. The dream of the West is the dream of the New World extended into the present.

I once said in print that the West – and I mean the interior West of plains, mountains, and deserts – is the geography of hope, the native home of optimism, the youngest and freshest of America's regions, magnificently endowed and with the chance to become something unprecedented and unmatched in the world.

I was probably demonstrating my thesis while I expressed it. I was shaped by the West and have lived most of my life in it, and nothing would gratify me more than to see it, in all its subregions and subcultures, both prosperous and environmentally healthy, with a civilization to match its scenery. When ever I return to the Rocky Mountain states where I am most at home, my native enthusiasm overcomes me, and I respond as unthinkingly as a salmon that swims past a river mouth and tastes the waters of its birth and feels an irresistible impulse, born I am sure of love, to turn inland in search of the stream where it was hatched.

But when I am thinking instead of throbbing, I remember what history and experience have taught me about the West's past, and what my eyes and nose and ears tell me about the West's present, and I become more cautious about the West's future. So I curb my enthusiasms, I begin to quibble and qualify; I say yes, the West is the native home of hope, but there are varieties and degrees of hope and the wong kind, in excessive amounts, goes with human disappointment and environmental damage as bust goes with boom.

–Wallace Stegner

INTRODUCTION

BY JIM CARRIER

All rivers rise with hope, and most river stories begin at the beginning, with a trickle of water high in the mountains. This one begins at the end. For all its promise and water, the Colorado runs out of both.

The man the Cucapa call *"El Coyote"* pushed his blue rowboat away from the tules, stroked through the murky brown residue of the Colorado River, and began to pull in his gill net. Halfway through the 40-meter net an 8-inch mullet landed with a lonely plunk in an orange pail at his feet. Minutes passed before I heard the plastic thud again, and by the time the whole net lay empty in the boat there were just three fish in his bucket.

"At least it's breakfast," Ricardo Sandoval said to me in Spanish. "The net's been out here two days." He paddled back to El Mayor, the ramshackle village of the Cucapa, the "river people" who are last in line for the waters of the Colorado River. But here, 50 miles south of the U.S. border, in Mexico's Baja California, the great river of the West is a shallow, narrow sump of salt and pesticide-laced crop runoff.

"Es nuestro vida — It is our life," said El Coyote, summing up 2,000 years of sustenance from the river. But for half a century the delta had been dying, and with it the Cucapa culture. No longer can tribal members hunt muletail deer, plant squash with the floods, harvest salt wheat, or eat fish three times a day. Several species of fish and plant life have disappeared. The Cucapa village has shrunk to 85 families, from a flourishing population of 1,200 a century ago. The once-rich estuary is filled with weeds and piles of trash, smoldering fires and occasional swamps of unhealthy water — barely enough to float their boats. Too salty to grow crops, too scant to support wildlife and too poisonous to bathe in or drink, the water that seeps by the village cannot in good conscience be called a river. In 1990, the fourth year of drought in the Colorado River basin, the water dropped to its lowest level in tribal memory. The Cucapa were lucky to eat fish once a week.

Sandoval pointed to a faded green rowboat lying forlornly in a dusty cracked path — once part of the river channel. "Waiting for water," he chuckled. "We are the river people. We are still here. But what river? It doesn't get this far."

While most maps draw a vibrant blue line from the Rocky Mountains to the Gulf of California, the nets of the Cucapa tell a poignant truth about the Colorado River: Its capacity to support the Southwest has finally been exceeded. For the first time in history there is not enough water to go around.

Plunging from frozen heights of 14,000 feet on the continental spine, the Colorado writhes 1,450 miles through mile-deep granite canyons to a sluggish desert sink 235 feet below sea level. It etches the Rocky Mountains. It carves the Grand Canyon. It deposits deserts by the sea. For only 56 years have its fabled, red-mud floods been under control.

European explorers thought the land it flowed through was useless. "Ours has been the first and will doubtless be the last party of whites to visit this profitless locality," wrote Lt. Joseph C. Ives of the Army Engineers in 1858, after steaming upriver past the Cucapa to the present site of Hoover Dam in search of a navigable route between the Rockies and the Pacific. "The Colorado, along the greater portion of its lonely and majestic way, shall be forever unvisited and undisturbed." As rivers are

Luring Las Vegas gamblers away from competing casinos, the Mirage Hotel's "Volcano" erupts four times an hour, sending gas-fed flames cascading down the waterfalls and into the surrounding lake filled with recycled water.

measured, the Colorado has little to brag about: Its 14,000-foot drop is the greatest in North America. It is one of the siltiest, carrying a pre-dam load of 380,000 tons a day. It is one of the saltiest, carrying nine million tons a year. And for sheer magnificence, especially in its canyons, the river has few equals. But it ranks only seventh in length and its water volume of 15 million acre-feet yearly is a pittance compared to the Columbia's 192 million and the Mississippi's 400 million.

And yet the Colorado binds the Southwest in a semiarid 244,000-square mile drainage (an area the size of France) and divides the region as no other element: state against state, farm versus city, Indian versus white. It has earned the reputation as the most legislated, litigated and debated river in the world.

In two years of tracing the Colorado I was stunned by what it is asked to do. The Colorado grows grapes in New Mexico, brews beer in Colorado, raises minnows in Utah, floats rafts in Arizona, lights jackpots in Nevada, nurses elk in Wyoming, freezes ice for California, and sweetens cantaloupes in Mexico. In bringing life to 25 million people and more than two million acres of farmland in seven states and two countries, it has reached a dammed and diverted denouement.

There is only so much water, and demands are increasing. Conflicts are constant among water users. The 1922 Colorado River Compact that divides its waters into two basins for use by seven bordering states — Wyoming, Utah, Colorado, New Mexico, Arizona, Nevada and California — seems inadequate as Americans flood the Sunbelt. New rules must be written in a time of environmental concern and heightened awareness of Native American rights and claims. As the Colorado River nears the end of a fruitful century in which it was harnessed to human needs, it enters an historic era of limits.

CHAPTER ONE

THE HEADWATERS

"These glaciers really did a number on this country," said John Barlow, a Wyoming rancher and rock 'n' roll lyricist, as we looked out the window of a small plane, which nearly touched the gouged granite of Gannett Peak. Sunrise had just topped Wyoming's Wind River Range, one of the Colorado's cradles. Down in the shadows little lakes by the hundreds lay frozen in July. I could see Peak Lake, Stroud and Mammoth glaciers, gray in color with stretch marks as they slowly ebbed into a brook. Like veins on a leaf, other brooks joined in to create the Green River, the northern reach of the Colorado system.

"You could say that the Green was the central river in settling the West," said Barlow. Some argue that the Green should be called the Colorado because of its length, 730 miles; its drainage, two-thirds of the river's total; and its history. It was the heart of the beaver trade, and famous rendezvous were held on its banks. The great pioneer trails, the Oregon and Mormon, crossed the river. John Wesley Powell began his historic exploration of the Colorado in 1869 at the town of Green River, Wyoming.

The Green also is unique in its wild beginnings, in the Bridger Wilderness Area. It becomes a forceful river within a few miles, unlike most other Colorado tributaries that are diverted in their headwaters. By the time it tumbles to Three Forks Park — a wet wooded basin — the river has a head of steam. It moves urgently, numbing cold. The deep grass is splashed with Queen Anne's lace, dandelion, and yellow and blue buttercups. Snow on the dark gray cliffsides never completely disappears, even in a drought year.

"This valley is what is good for our business," said Terry Reach, a wilderness outfitter who takes clients into the headwaters by horseback. "This is one of the few places where there are so many things to do: hunt, fish, watch wildlife." As we rode through the valley, the sound of water created a stereo effect: Porcupine Creek on the right and Marten Creek on the left. We passed through remnants of an old forest fire, its blackened trunks still standing. But the sidehill is ablaze with bluebells, paintbrush and wild rose.

The river slows and begins to meander, pouring its silt into Upper Green Lake. I could see two moose, a cow and her calf, clomping at the edge. The lake is brilliant turquoise from reflected glacial silt. Two miles more and the river leaves the wilderness, a flat blue serpent through a wide valley of sage. It buttonhooks and heads south. This is where its virginity ends. The first irrigation ditch, the Canyon Ditch, cuts into the river's bank and, following the contours of the land, slowly begins to run away through the sage, toward hay pastures on nearby ranches. When water is spread onto sage, the sage dies and grass for cattle grows in its place.

Over a ridge, on the New Fork of the Green, John Barlow's grandfather, Perry W. Jenkins, a Columbia University mathematician and astronomer, built his ranch in 1905. He created a brand from a plus and minus sign and called it the Bar Cross. He also organized Sublette County around the Green's watershed and formed an irrigation district to grow hay and cattle. Later he helped represent Wyoming in negotiations for the historic 1922 Colorado River Compact.

"He was into water," said the bearded, Levi-ed Barlow, who inherited the ranch after the death of his father, Norman, a state senator. In the 1970s John fought the construction of a dam on the Green near the ranch. The water would have gone to mine coal. "That's when my father realized I didn't have that beaver mentality."

Each summer John flood-irrigated 2,200 acres. Over several months he poured eight feet of water on his fields to grow grass for 1,100 cows. That's a lot of water for one cutting of hay but ranchers say they need to raise the water table 8 to 10 feet before it reaches the roots of the grass. By filling the underground aquifer they create wetlands used by blue herons, sandhill cranes, Canada geese and ducks. One study also shows the hayfields act like a reservoir, slowly releasing water into the streams and rivers year-round, like the glaciers within sight of the ranch. It is said that because of such "return flows" on the Colorado River, every molecule of water is used three times.

In the late 1980s, John Barlow lost the ranch to high debt and low beef prices. The new, absentee owners sold the cattle. The unused water slipped into the Green, where almost 60 percent of Wyoming's compact share goes for lack of use. Other wealthy neighbors who have bought up ranches have turned their irrigation ditches into trout streams and replicas of European babbling brooks. Barlow fears losing water to the Sunbelt — by default or by new marketing proposals to lease ranch water to developers in Arizona, California and Nevada. "I would like my kids' kids to live here," said Barlow, who makes a living writing songs for the Grateful Dead. "I don't think the national interest is served by running all the water to where it cleans off driveways in Los Angeles. But I think the 1922 interstate compact that carefully divides the water between states will be abrogated, and Wyoming will be the loser. This is heretical but I don't know how we can justify our need for the water under present circumstances."

Like looking for a needle in a haystack of minnows. U.S. Fish and Wildlife specialist Bruce Haines seines the Green River in Dinosaur National Park in search of the larvae of the endangered Colorado squawfish.

Three hundred miles to the southeast on the Continental Divide in the state of Colorado, the river has another beginning, on the breasts of the 12,000-foot Never Summer Mountains. But here, in Rocky Mountain National Park, the birth is painful: The runoff of the serrated mountains is interrupted by a gouge, a 10-foot-wide ditch that runs 14 miles across the mountains. The Grand Ditch runs water eastward across the divide at 10,186 feet, then sends it down the east face of the Rockies to Fort Collins, Colorado and 30,000 acres of sugar beets, corn and barley on the Great Plains.

"Some people thought it was awful to tear up the side of the mountain," said Harvey Johnson, 95, chairman of Water Supply and Storage Company, which owns the water in the ditch. "I tell them we're growing food, and they'd go hungry without it."

Dug by Chinese laborers, the ditch carried water by 1900. "We were quite desperate and the Western Slope was flush with water," Johnson told me as we walked the ditch's banks. "The ditch company decided they'd just go get it." That was the mentality and legal right in the West. The primary water law, "first in time, first in right," gives the oldest users of water nearly ironclad seniority and ownership. Johnson, carried to Colorado in a covered wagon, spent his life making the semiarid plains bloom. "It's very productive soil if you put good water on it."

Colorado's entire Front Range, the land that lies at the eastern foot of the Rocky Mountains, is a rich farm belt and growing metropolis because of water diverted across the Continental Divide. The Grand Ditch is the oldest, but there are 20 others, draining a third of the Colorado's high tributary flows. Skiers at Winter Park race downhill next to huge green tubes that carry snowmelt to Denver, a city that gets 55 percent of its water from the Colorado River. Deeper snow means better skiing as well as greater runoff in the spring.

The most improbable diversion lies 2,000 feet below the Grand Ditch — the Colorado-Big Thompson Project. A federal engineering marvel, the project collects snowmelt in a reservoir called Lake Granby. Huge pumps literally lift and push the water up the old Colorado riverbed, to Grand Lake, a natural lake.

Resembling a checkerboard on a mountain, the Fraser Experimental Forest is really a grand experiment in water farming.

Below the surface of Grand Lake a huge pipe drains the water. A beautiful, natural mountain lake is thus made part of a plumbing system that takes 90 percent of the fledgling main stem's water. A tunnel 13 miles long and nearly 10 feet wide, also part of the Colorado-Big Thompson Project, takes the lake water under the Continental Divide to the east face of the Rockies. The water tumbles downward, turning five turbines that generate the power needed to push the water from Lake Granby. The water then flows to Front Range cities, including Boulder, with a combined population of 500,000, and to Weld County, the fourth richest agricultural county in the United States.

While all this is going on, each summer yachts are racing back and forth on the 4 1/2 mile Grand Lake in E-skows. "When in doubt let it out," yelled veterinarian Jim Munn at his crew as we skimmed the lake. Ninety members of the Grand Lake Yacht Club, the world's highest registered yacht anchorage, compete for the coveted Lipton Cup, an ornate solid silver cup first presented in 1912 by yachtsman and tea magnate Sir Thomas Lipton. Club members know the cup is priceless, and winners have taken odd measures to safeguard it during their year's reign. Some winners have kept it under their beds at home. A gust of wind tipped us wildly and we hiked out to steady her, just as Munn yelled: "We're coming up on Hurricane Alley" The fluky mountain winds make yachting here a unique experience.

When the Colorado-Big Thompson Project was first engineered, Grand Lake would have risen and fallen according to pumping needs. Absolutely not, said the old-money cottage owners and Yacht Club members. Their clout guaranteed that the level of the lake remained at their shingled cottage doors.

The river that leaves the project through the Lake Granby dam is a small stream in the mountains, with just enough water to meet the state requirement for keeping trout alive. "Without that requirement you would have dry streambeds on the Western Slope for sure," said Rolly Fischer of the Colorado River Water Conservation District, which has carried on a half-century water war with Denver and the Front Range. Formed as a "protective association" when the Colorado-Big Thompson was built, the district has fought nearly every transmountain diversion. "The fear has been that the Western Slope would be dewatered just as California's Owens Valley was dried up by Los Angeles," said Fischer.

Most of Colorado's people live on the eastern slope, and almost all the available water is on the Western Slope. Aurora, Denver's suburban neighbor, proposed tapping the Gunnison River, the one remaining Western Slope river not diverted to the Front Range. "The Gunnison is one of the last frontiers in the water wars," said Bill Trampe, one of the ranchers near the Gunnison who increased their own taxes to fight Aurora in water court. "Recreation is the Gunnison's leading industry" he said. "It requires water in the streams. And half of Aurora's water would go on lawns." Aurora's population of 222,000 could triple by the year 2050. The city is already using the water from 20,000 acres of mountain ranchland and has bought other farm water. "The Western Slope views it as their water, while in reality state law provides for diverting water to where it can be used," said Tom Griswold, Aurora's utilities director.

In the mountains of the Colorado watershed, snow depths are watched like a water stock market. Each winter month, government surveys measure snowfall to predict runoff. "I can tell you right now it isn't going to compare very well to a normal year," said Chuck Troendle, a hydrologist with the U.S. Forest Service as we skied through the Fraser Experimental Forest west of Denver measuring snow. Along a blaze-marked course at 10,600 feet we stopped periodically to ram an aluminum pipe into the snow. As he measured the depth and weighed the water the figures were disappointing. "This watershed should average 12 inches of water. This is 7. In some places it's 2 or 3 inches."

It was April 1, and water was beginning to flow in the streams below. While there appeared to be plenty of snow in the mountains, and water in the creek beds, the Colorado River system was entering another year of official drought. This would be the fourth year of drought for those who depended on the river's water. As we traversed through an old growth lodge pole stand, the sun warmed us enough to shed a sweater layer.

"This is the source of water for the western United States. Fifteen percent of the land mass produces 85 percent of the water in Colorado," said Troendle, a big amiable scientist in an orange parka and jeans. "The amount of water in the snow today will

The Ouray National Wildlife Refuge uses man-made ditches, Green River water and nesting islands in its marshes to create habitat for waterfowl.

account for 70 percent of the total annual flow. Another 5 to 10 percent will come as rain or snow from now until June 1. The rest falls as rain through the summer and fall. This is the reason we monitor on April 1. You can forecast pretty closely what stream flow will be."

Troendle's studies show that the river's flow could be augmented by selective clear-cutting of trees, especially if combined with cloud seeding. A spruce-fir canopy, for example, can stop 25 to 50 percent of the snow from reaching the ground because the snow evaporates at tree-top level. If trees are logged in an open cut, runoff increases by 50 percent on a north-facing slope. In times of drought there are renewed calls to cut trees to augment water flow, not only in the high country, but in cottonwood groves that line the river for hundreds of miles. But clear-cutting ignores the role of standing timber as animal shelter and in holding water in the soil.

The snow survey results from the Fraser forest and 1,500 other spots on the Colorado's sprawling drainage are fed into computers in Portland, Oregon by the federal Soil Conservation Service, which forecasts river flow. The data is used to regulate reservoirs, plan power generation, order seeds and determine the health of fish and wildlife. Once the water leaves the mountains, it enters an arid area.

"Every drop counts when you run a river through a desert," said David Johnson, snow survey program manager for the Soil Conservation Service. "Sitting here we see a precarious position between supply and demand. The trend is increasing demand. Every year there is only so much." But as the drought continued, Johnson predicted the response in the West would be the same: "Tighten belts and get through it. It's like safe sex. The more successful we are at getting through the drought, the more you encourage growth. I don't want to be anti-development, but people don't see the precarious balance we are living. Conservation should be practiced continuously and religiously. We just aren't real serious about it. We're setting ourselves up for such an enormous hurt."

Punching commands into his computer, assistant Garry Schaefer was able to tell me that the snowpack for the Colorado basin was 30 percent below normal, and predicted that but because of four years of lean snowpacks, the actual water runoff would be a whopping 70 percent below normal. That would reverberate all down the river.

Graced by the reflection of a passing cloud, a beaver pond near the abandoned gold rush camp of Lulu City in Colorado's Rocky Mountain National Park is the first impoundment on the Colorado River. In the calm evening deer visiting the pond make it easy to forget the turmoil these waters will endure in their futile march to the Pacific Ocean.

Right

Charlie Settlemeyer fixes the flag on his ranch overlooking the Colorado River near Bond, Colorado. The river blocks his road to town so Settlemeyer regularly drives his all-terrain vehicle across the Rio Grande Railroad bridge with his dog riding on the handlebars.

Far Right

Jim Taussig irrigates the lush grass carpeting his ranch near Kremmling, Colorado. This was all sage brush before Taussig's father diverted the Williams Fork River early this century to create pasture for his cattle. Taussig is out every day with his shovel keeping narrow canals open to flood the hillsides evenly. But the water belongs to Denver. His father sold the water rights in 1963 and got a 40-year lease on the water.

Frank Norton on the site of his family homestead, now the lakebed of Lake Granby in central Colorado. The lake is a catchment basin for the Colorado-Big Thompson water project that diverts water under the Continental Divide to the Front Range. When he was a boy Norton helped his father on the family's Circle "H" Dude Ranch, seen in the photo he holds. Lake Granby was filled in 1952, and the Norton family moved up the hillside to open a marina.

Beth Munn battles the tricky mountain winds during the annual Lipton Cup Regatta on Colorado's Grand Lake. Surrounded by expensive summer homes, Grand Lake once styled itself as the headwaters of the Colorado River. But now the water runs the other way, from Granby Lake through Grand Lake to the entrance of the Adams Tunnel at the lake's east end and on to the Front Range.

Loaded with sediment from the Uintah Mountains, the Green River took millions of years to cut more than 3,000 feet down into the Split Mountain anticline in Dinosaur National Monument.

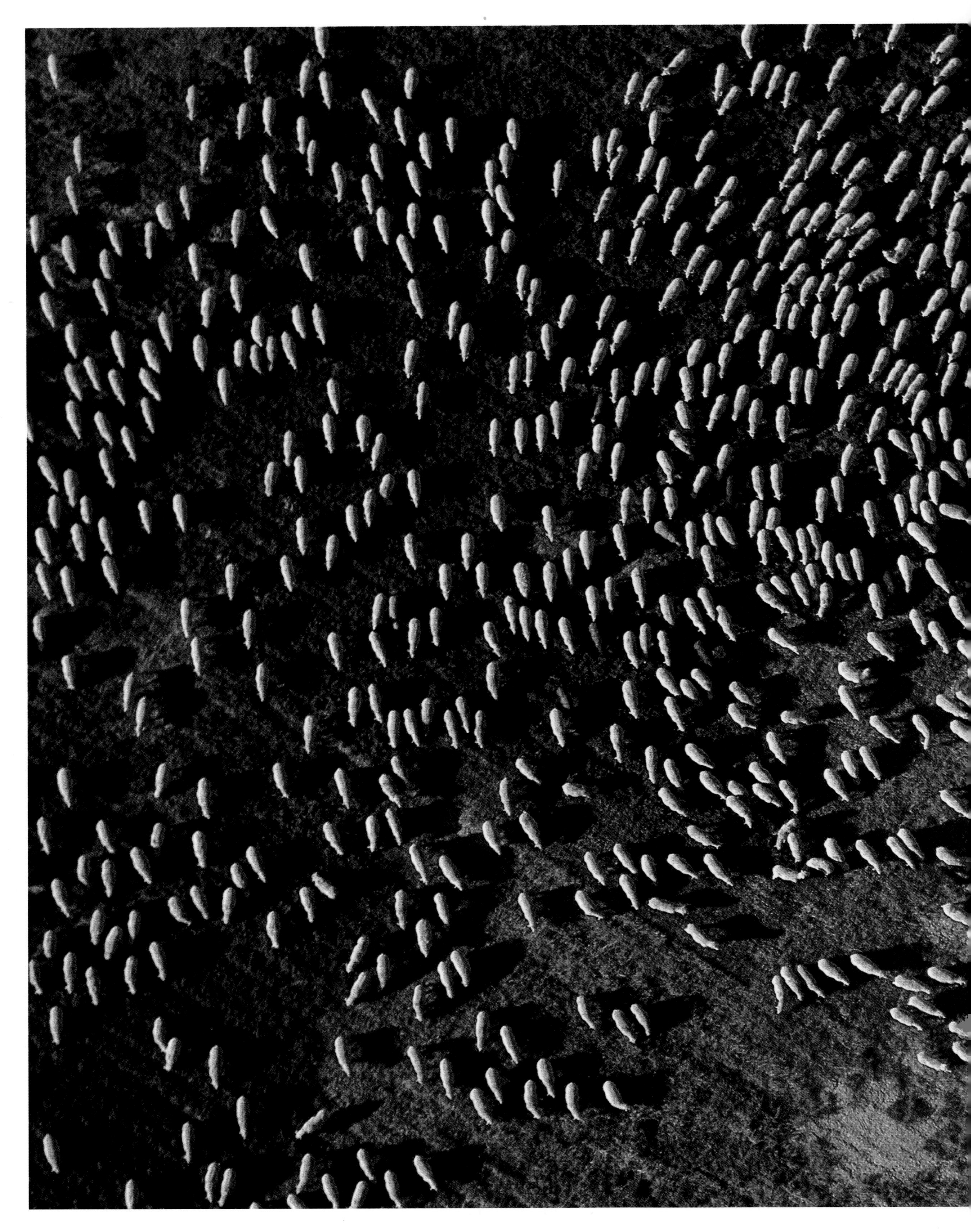

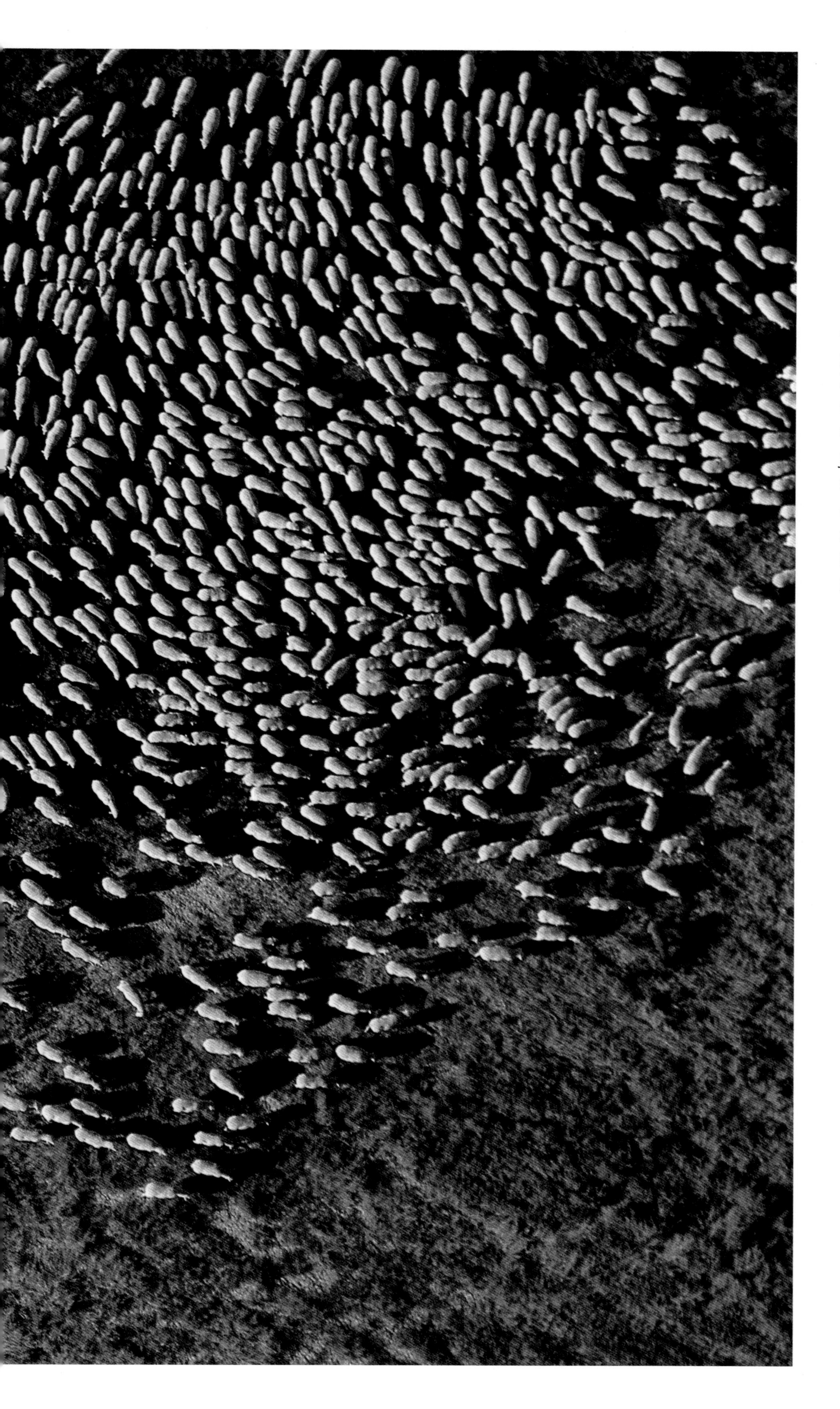

Sheep graze on irrigated pasture near the Green River in Utah. Livestock uses more water in the Colorado Basin, in the form of pasture and hay, than all human use combined. Even during the recent drought years, the biggest use of Colorado River water is agriculture.

Above

Out for the annual snow survey on the Fraser Experimental Forest, Colorado State University researcher Jim Meiman steps carefully across a snow pack bridge on the Fool Creek Drainage. The survey has measured the snowpack for the last 30 years to determine the effect of selective tree cutting on water production. Water from this snow belongs to Denver, and some of it actually goes to make snow at a near-by ski area.

Right

People steal water in the arid West, so rancher Jim Taussig puts his under lock and key. Taussig remembers the times, before his father built this headgate, when every spring they cut a cottonwood tree and wedged it across the Williams Fork river to divert the water to their meadow. Now Denver owns the water, and Taussig's lease on it runs out in 2003.

Overleaf

Broken Foot creek spills down the side of Rocky Mountain National Park, but the Grand Ditch, the first diversion of the Colorado River, diverts it across the Continental Divide before it reaches the Colorado River below. The Grand Ditch runs for 14 miles across the face of the Never Summer Range, diverting 20,000 acre feet of water to the Colorado Front Range annually.

LEFT
Deep beneath the the Continental Divide, Bureau of Reclamation managers stop on an inspection trip through the 13.2 mile long Alva B. Adams tunnel to paint their names on the ceiling, a Bureau tradition. The tunnel, part of the Colorado-Big Thompson Project that carries water from Grand Lake to Estes Park for Front Range water users, is drained every three years for a safety inspection.

ABOVE
Every spring finds Fred Loeskan in the lonely reaches of Rocky Mountain National Park cleaning snow and rockslides out of the Grand Ditch. The Grand Ditch was built by Chinese laborers before Rocky Mountain National Park was created. The farmers own the water because they claimed and developed it first.

OVERLEAF
Shaped over millennia by the Green River, Steamboat Rock stands like a monument in Echo Park, part of Dinosaur National Monument. In the 1950s a proposed hydro-electric dam just downstream would have turned it into a modest island. In a compromise, environmental groups dropped their opposition to the Glen Canyon Dam in Arizona in exchange for saving Echo Park from reservoir waters.

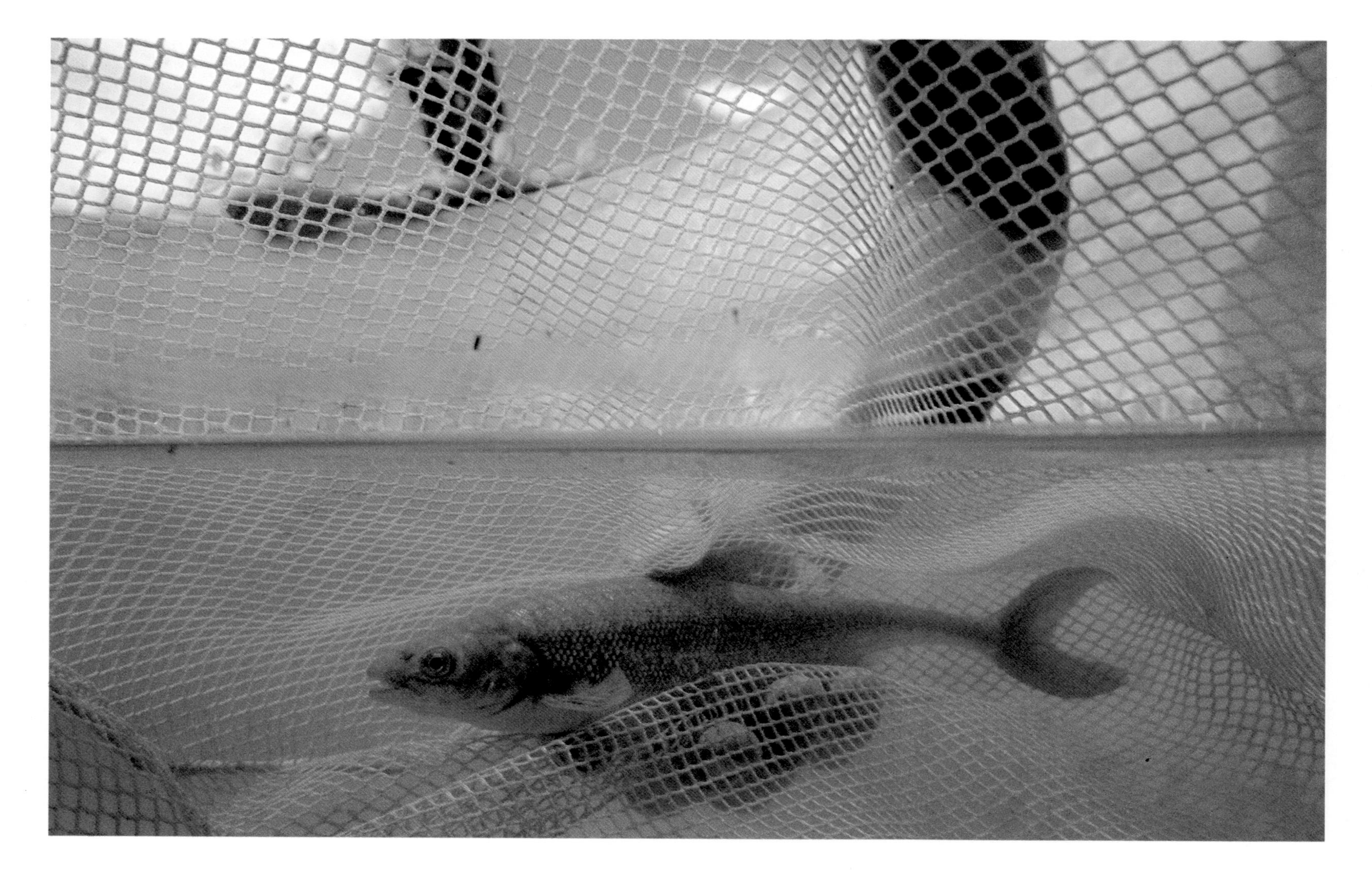

Endangered fish, like this Bonytail Chub, take special handling to survive in the wild. Fishery biologists at a U.S. Fish and Wildlife hatchery near Vernal, Utah, are trying to understand the conditions the fish need to spawn successfully. Some species have not spawned successfully in decades (perhaps since the building of the big dams), and the wild fish are thought to be very old — possibly too old to ever spawn again.

Back from a predawn seining expedition at the confluence of the Yampa and Green Rivers, researchers Eric Tsao and Mark Brady settle at a picnic table in Echo Park to look for the larvae of the endangered Razorback Sucker. Identifying the larval species is tough, even for experts, so their tiny catch is stored in bottles for later lab analysis.

In the fading light of sunset, researcher Mark Brady checks a fish trap on the Yampa River at Echo Park in Dinosaur National Monument. Brady is one of many researchers trying to determine spawning patterns and habitat requirements of endangered fish. The Yampa River, relatively undammed, runs cold and wild during the spring runoff and warm and sluggish in the summer, making it, perhaps, the best remaining hope for saving the fish.

CHAPTER TWO

THE CANYONLANDS

From the mountains the two arms of the great river, the Colorado and the Green, flow unperturbed — lonely almost — through a landscape that becomes increasingly arid. Precipitation drops from as much as 55 inches per year in the mountains to as little as 10 on the Colorado Plateau, the vast plain west of the Rockies in Colorado, Wyoming, Utah and Arizona where the rivers sink into serpentine cracks. This is the start of Canyonlands, renowned for white water rafting, spectacular scenery and heated environmental debates. In spite of the drought, both rivers carry plenty of water here, in part because the upper basin states of Colorado, Utah and Wyoming do not use their full share of the river. Colorado, for example, allows one-third of its compact share to flow out of state. But most of the water belongs to the downstream states of California and Arizona and Mexico.

Because of the deep canyons, the water is not easily used for irrigation, but the precipitous walls make ideal dam sites. Many more dams have been proposed than built. An area called Rattlesnake Canyon near the Colorado-Utah border was picked as a hydroelectric dam site by a private company, but later abandoned. It would have flooded one of the prettiest stretches on the river, and been upstream from Westwater, a favorite white-water run. It was on the Green, before Flaming Gorge was flooded, that the modern sport of river running began in 1891. The flat-bottomed Galloway boat, with uplifted fore and aft, revolutionized the hazardous trips by facing the boatman downstream. Nathaniel Galloway made several trips through Flaming Gorge, past Vernal and at least once to the confluence with the Grand River, as the Colorado was known.

The canyons also meant the Colorado never attracted a major city to its banks. With no major cities, pollution is not a problem. Grand Junction, Colorado. contributes some radioactivity from an old uranium tailings pile, but that is being cleaned up. The greatest pollution is naturally occurring salt. Starting with snowmelt quality of 50 milligrams of salt per liter, both the Green and the Colorado systems grow increasingly saline as the water is extracted, evaporated from reservoirs, passed over natural salt beds and poured through soil that was once the bottom of an ancient sea. By the time the Green reaches Green River, Utah, 600 miles from its source, its salinity exceeds the federal government's 500 parts per million standard for drinking water.

In the Grand Valley of Colorado, where the Bureau of Reclamation has begun a multimillion dollar program to line canals with plastic and concrete to reduce salinity, 70,000 irrigated acres add 580,000 tons of salt to the river each year from runoff and soil leaching.

Before man's development of the river, the Colorado used to flood a silty cold in the spring and trickle warm and clear in the fall. Now, below dams like Flaming Gorge, it runs clear and cold year-round because the dams trap silt. Reservoirs release frigid water from their depths back into the river, creating excellent trout habitat, but contributing to the near extinction of several native species.

"We're not talking about one or two fish. We're talking about the loss of an entire fauna," said Harold Tyus, who runs the U.S. Fish and Wildlife Service laboratory on the Green River near Vernal, Utah. "More than half of the endemic fish are listed as endangered. This is the last stronghold of the squawfish. There are more here than anywhere else in the world. It is the only known spawning ground."

Before sunrise in a rose-colored canyon I watched Tyus's biologists seine the Green for tiny Colorado squawfish larvae — "threads with eyeballs." For some reason, the greatest drift of squawfish larvae is at night and dawn. Perhaps this is an adaptation to avoid predators. Diane Moses waded into the cool river with a 15-foot drift net. "This is where I really wake up," she joked. The water was 19 degrees and she was in sneakers and shorts. She hooked the net to rebar stuck in the muddy water, and retrieved it 90 minutes later. She scraped the debris in a Zip-lok bag, and added alcohol, to tint the fish. One showed up. The fish once grew to monster size, six feet and 100 pounds. One bumped Major Powell's boat and was thought to be a rock or whale. When settlers first came to the Green, they hung squawfish in trees to strip off their white salmon flesh. Historic photographs show bony skeletons swinging from branches. The fish were eaten for Christmas dinner and at July 4 celebrations.

That they exist at all today is due to the Yampa River, the only

Swimmers linked foot-in-arm enjoy the turquoise waters and shallow rapids of the Little Colorado River near its confluence with the Colorado in the Grand Canyon. Dissolved travertine gives the water its beautiful, unreal color.

major undammed tributary in the Colorado drainage. It joins the Green in Dinosaur National Monument and runs warm and low enough for squawfish spawning. Proposals to dam the Yampa — one company wanted to sell the water to San Diego — would probably eliminate the fish, said Tyus, who favors giving the fish a water right of their own.

The endangered fish — the squawfish, the humpback chub and razorback sucker — have been labeled "junk fish" by sport fishermen who have introduced trout, bass and other exotics to the Colorado system. Those who labor to save the endangered fish have little public support. "If you go back in history, you'll find these fish were very abundant," said Tyus. "Razorbacks were taken out by the thousands, salted and carried in wagons to sell." There are now perhaps 1,000 razorbacks left.

But the endangered species are having considerable impact on traditional management of the river. Enough water must now be released from tributary dams as far away as Gunnison and Aspen to keep the fish alive. The releases must be manipulated for temperature and flow-force to approximate the once natural flow. Flaming Gorge, already modified once to raise the temperature of its releases, may have to change its peak power releases to help the fish.

At Vernal, Utah, where the Green spills out of Split Mountain, Mormons turned the wide valley a productive green. But below the farming community the river corridor narrows again through the pink and brown eroded buttes of the plateau. It makes slow turns through Desolation and Gray Canyons. The river takes its sweet time, cutting, eroding, flowing through bow-knot bends of red rock. Below Green River, Utah, the land rises again and the river sinks deep to join the Colorado.

Where the two meet, in the remote labyrinth of Canyonlands National Park near Moab, Utah, a powerful river results. I could see the pistachio color of one mixing with the red-silt-laden flow of the other, a total of 13,000 cubic feet a second swirling and gathering strength for a plunge through Cataract Canyon. Along the river the life zone is narrow, a green mat for the chocolate water. "Water attracts living things," said Tex McClatchy of Moab, who runs jetboats to the confluence. "What makes the Colorado so magic is, we see deer, we hear coyotes practically every night. I am one who has not connected to another society. My days like to run loose. There's no fences. I can go anywhere I want to. The river, to us, is free flowing life."

The land tightens around the river and lifts — and the water becomes rapids. The river drops one foot for every mile here. In a

In Canyonlands National Park, the Green River (right) and the Colorado River (bottom) join forces at Confluence Park.

lifejacket and rubber raft I joined a group of thrill seekers through Big Drops 1,2 and 3 — rolling, muddy rapids guaranteed to soak to the skin.

Around us in the slickrock loomed some of the West's hottest environmental battles, brought to life in Edward Abbey's novel, *The Monkeywrench Gang.* The real life issues pit mineral extraction and what Abbey called "industrial tourism" against the wilderness designation.

"Where environmentalists stopped everything in southern Utah they have half of the population of 50 years ago and the highest unemployment in Utah. That's what tourism does for you," said Cal Black, an ex-uranium miner and San Jose County commissioner, who, before he died, proposed bringing a nuclear waste dump to land adjacent to Canyonlands. Abbey used Black as nemesis Bishop Love in his book. Said Black: "You can have all the beauty in the world, but if you can't have development or private land, you'll have practically no benefit from it. You can't go anywhere in Utah and buy a home with a view of Lake Powell, even though 97 percent of the lake is in Utah."

Black's frequent opponent, Ken Sleight, an outfitter in Moab, likes the river just as it is. "The river is not populated. It has helped very few people — except downstream users. I'm one who says let 'em have it. We still have a river. Why transplant population from one place to another? Why not feed them downstream?" Sleight, the mold for Seldom Seen Smith in the *Monkeywrench Gang*, concludes: "I once believed in dams. I was a farmer. But dams are built mainly to promote power, and we destroy great rivers to make money."

As we floated through Cataract, mysterious side canyons beckoned with carved stone structures worthy of national monument status themselves. These are places rarely visited, places of imagination. We poked through the ruins of Anasazi Indians, the "ancient ones" who lived along the river a thousand years ago. The best explanation for their disappearance was a combination of drought, overpopulation and internal strife — elements also present in the West today.

"The Colorado is a case history on the economics of the West," said U.S. Senator Tim Wirth of Colorado, who helped organize the float trip. "The West is a United States colony," he said, a reference to taxpayer-subsidized water projects that have spread the Colorado's waters to irrigation projects that build cities in deserts and grow surplus and subsidized crops with precious water.

"The Colorado has defined what the West has become," the rafters heard from Sally Ranney, president of American Wildlands, a national environmental group based in Denver. "And because of the water shortage it will define what the West will not become. We have a saying out here that water flows toward money. It has nothing to do with gravity."

The anxious river, having just begun its full surge, started to eddy near the end of our trip. Our rafts grounded on hidden sandbars, and walls of sand rose on either side to choke side canyons. The collected silt of two great rivers, which used to flow to the sea, has begun to clog the upper end of Lake Powell, causing a problem. "In the old days there were rapids to Hite, Utah. The biggest was at Dark Canyon," said Sherri Griffith, who commanded the rafters. "They said it would be a couple hundred years before it silted in, and this is how much it silted in, in 25 years." As we would see farther down the river, the problem with the silt deposit in Lake Powell was the tip of the iceberg.

Capable of holding nearly two years' flow of the Colorado in a red sandstone bathtub 1,900 miles around, Lake Powell is about halfway down the river. Just below Glen Canyon dam, Lees Ferry is the official dividing point between the upper and lower basins. The 1922 compact apportioned about half of the flow to each basin, or 7.5 million acre feet, from an annual flow then thought to average about 17 million acre feet.

Virtually all the water that will enter the Colorado has done so by now, and the lake is a quick look at supply and demand. In 1990, after four years of drought, a ring was showing in the bathtub 66 feet above the water level (evaporation alone takes five feet a year). According to the compact, in years of drought, the lower basin of California, Arizona and Nevada gets its share first. Less than 5.5 million acre feet of water flowed into Lake Powell in 1990, not nearly the 8.25 million required downstream by the compact and Mexico.

Lake Powell's creation in 1963 was the crowning act of the Bureau of Reclamation's 30-year, big-dam era. Built for water

Larry Creech died of a heart attack while visiting Lake Powell. A year later his family scatters his ashes into the deep waters of the lake.

storage, flood control and power, Glen Canyon Dam flooded caverns and canyons that only a few thousand people had ever seen. Today more than three million people visit the vast desert lake each year. Hundreds of houseboats ply the water spring through fall, stopping each night in the nooks in the rocks. The dam and lake were a compromise, traded, essentially, for proposed dam sites in the Grand Canyon below, and Dinosaur National Park above. Glen Canyon also is the keystone in developing other dams and irrigation projects in the upper basin. Its hydropower pays the bills. The dam also holds a drought cushion.

But rising by the dam is a new conflict on the river, as the traditional water users — the irrigators and power interests — bump heads with the enormous economic force of leisure time. Boaters want their docks in the water, not draped on silt. The power brokers, who make $72 million from Glen Canyon's hydroturbines, cannot ignore them. Having created a marine paradise, its hard to ignore the users.

Down river, in the Grand Canyon the dam was exacting another price. With most of the river's silt blocked off, the clear, deep-green "hungry" water eats away existing sand and silt, the base for the canyon's ecosystem. "The worst effects of the dam came early, in 1964," said Martin Litton as he maneuvered a dory through the rapids below the dam for the 75th time in his 74 years. The Grand Canyon curmudgeon has long been a burr under the saddle of dam interests.

Litton had rowed and explored the canyon before the dam gates closed; his first trip was in 1955 when he and his wife were the 185th and 186th people to run the Grand Canyon.

"You'd see dead cows, railroad ties, telephone poles, 50-gallon drums. There was so much driftwood. Places we think of as good camps now, you wouldn't think of using before. We slept on wet sand, so it wouldn't blow in our ears. We slept close to the river, to keep cool." He noted a line of dead mesquite trees high on the banks — they used to get watered once a year by spring floods. "A dam site was surveyed every half mile on the Colorado in the Grand Canyon," proposals that would have flooded the scenic wonder. He pointed to drill holes halfway up the red-stained

In Potash, Utah, Colorado River water is used to wash potash ore from underground strata and carry it to settling ponds.

limestone. Proposed Marble Canyon Dam would have backed the water 50 miles to Glen Canyon. "We would have missed 40 miles of rapids." He rowed on, then added philosophically, "Glen Canyon dam will be a temporary aberration. It won't be here in another 100,000 years."

Without the usual feast-and-famine flows of the natural river, wildlife changed abruptly. Fish used to warmer waters and muddy bottoms died off. Beaver disappeared because entrances to their homes, built underwater in the riverbanks, were regularly exposed as the water level rose and fell. Tamarisk invaded and songbirds followed. Trout were introduced, and bald eagles began to make winter stopovers. Peregrine falcon, another endangered species, showed up to chew on songbirds. Litton glanced up at the violet-green swallows looping about for bugs: "They're doing fine, but most cliff swallows left after the water cleared. Not enough mud for their nests." With fewer beaches for rafters to camp on, the national park limited visitors to 22,000 a year, outlawed driftwood fires except in winter, and made everyone carry out all waste.

We stopped to see the wonder of green blooms in the hot desert: monkeyflowers and maidenhair ferns hanging around small waterfalls that flow into the river from side canyons. At Nautilus Canyon we climbed to see the fossils in the rock and Litton observed: "This was a wonderful beach. Now it's just a rockpile. There's very little bare sand left." Above us a canyon wren sang its haunting, sinking song.

On our second morning in the Canyon, we awoke to find the wood boats high and dry on the narrow beach. The water had receded nearly 13 feet during the night. Glen Canyon's hydroturbines are used when power demand peaks, causing the water in the Grand Canyon to go up and down like a tide. Less demand for power, less water. "See, the water is low today because it was cool in Phoenix yesterday, and they didn't want as much air-conditioning. The beaches can't take this daily up-and-down stuff," explained Litton, who argues for a shift of peaking power away from Glen Canyon.

The fluctuations keep boatmen constantly calculating their available water. With a schedule to keep they must time their departures to reach difficult rapids along with high water. The farther down the canyon they get, the more difficult the calculation becomes. But the greater problem is the ecosystem.

"We're in a losing situation," agreed Dave Wegner of the Bureau of Reclamation whose studies showed the dam's operation had "substantial adverse effects on downstream environmental and recreational resources. In this drought year there is not a lot of water so it is managed for maximum electricity.

"The Grand Canyon National Park wasn't a player when Glen Canyon was built," said Wegner. "It was a weak sister, with no weight in Washington." As a result, little regard was given to the effects of the dam on the great park. The situation worsened in 1977 when control of power shifted from the Department of Energy to the Western Area Power Administration, which began maximizing electric generation. "It was a subtle but significant shift."

It is true that demand for electricity controls the dam, said Lloyd Greiner, a manager with the Western Area Power Administration. "I don't believe there is enough evidence that fluctuating flows are a major contributor to the damage. The river drops 2,000 feet in the canyon. With water rushing through, there will be erosion." William Clagett, chief of Western Area Power, told Congress that changing the dam to constant releases would require less efficient, or environmentally more costly generators, like coal-fired turbines, to make up 850 megawatts lost by the lower-flows through Glen Canyon's turbines. He argued that power generation has primacy in operating the river.

The government's environmental studies also show that the dam has not been all bad. Vegetation has increased. An exceptional trout fishery has been created, and the white-water boating season actually has been extended by the dam.

While we were in the canyon, a congressional hearing was looking into the effects of the power surges and wrestling with proposals to limit the range of power surges. The Interior Department agreed to temporary limits while it undertook a major study of the canyon. American Rivers, an environmental group, named the Colorado the nation's most endangered river because of the changes in the Grand Canyon. The Bureau of Reclamation also began exploring possible mitigation steps, including one idea to "mine" silt from Lake Powell and dump it

Waste water from uranium operations at Atlas Minerals on the banks of the Colorado River west of Moab, Utah stands waiting for a cleanup operation to begin.

into Paria Canyon, near Lees Ferry, to allow it to replenish the canyon. One senator from Wyoming also proposed building a second, regulator dam, below Glen Canyon. Stewart Udall, who was a congressman when Glen Canyon was approved and saw it completed as interior secretary, said America's views changed as the dam was completed. "Most of us...learned that hidden canyons and free flowing rivers...stand at least equal on the American people's scale of values to monumental dams and cheap hydroelectric power," he told Congress. He said he was baffled by Western Area Power Administration's assertion that the law creating Glen Canyon Dam requires operation for maximum peaking power. He said Congress did not intend to operate the dam as a daily threat to the Grand Canyon.

While the arguments echoed off the Grand Canyon's walls, I tried to listen to other sounds, too: the cascading call of the canyon wren, the roar of a rapid around a bend, the tinkle of spring water through maidenhair ferns. They still are reasons to come. At the mouth of the Little Colorado, we stopped to play in the Caribbean-blue waters, caused by silt. We wrapped life preservers around us like diapers and slid feet-first down plumes of warm water. This one tributary also contains a few humpback chubs, one of the endangered species.

On a Sunday morning we awoke in the canyon to bad news: Low weekend demand for electricity in Phoenix meant that low "Saturday water" was reaching us 87 miles from the dam. In Unkar Rapids, Litton hit a rock. Midway through Neville Rapids another boatman struck a rock and put a hole in "Ticaboo." Two miles further the river looked worse. "Oh my God" said Litton, when he saw the boulders sticking out of Hance Rapid. "It's a hot day, that's in our favor. But it's Sunday, and that's not in our favor. I've never seen Hance this low. This is basically unrunnable in dories."

So we waited, with environmentalist Litton praying, ironically, that Phoenix would suffer a heat wave so we could float the river. That night I stared up into a heaven cut by the cleavage of the canyon, like an observatory. I watched satellites inch across the star-sprinkled sky and thought, mine was the last generation to see a night sky uncluttered by man-made things — or the Grand Canyon's water unregulated.

At dawn the water was even lower, so those of us who had to leave walked around Hance and boarded rubber rafts that could bump over the rocks. The dories, with half the crew and half the food and water, would wait for "Monday water" when Litton's prayers would be answered.

The seemingly indestructible solid rock landscape of the Southwest can become quite fragile under the pressure of millions of visitors. Water attracts people— more than three million a year to Lake Powell and Glen Canyon Dam. A popular overlook downstream shows the inevitible grafitti that comes with visitors.

WAYNE
TERRE
GRA
PAT

RIGHT

Green doglegs grace the red rocks of the cliffs above the Colorado River just downstream from Glen Canyon Dam. The municipal golf course of Page, Arizona, gets 125 million gallons of wastewater a year from the city that got its start with the building of the dam. Page, like many western desert cities, is sensitive to criticism of wasting precious water and uses effluent for what would otherwise be considered a wasteful luxury.

FAR RIGHT

Two acres of lawn sit at the bottom of Glen Canyon Dam, covering the roof of the penstocks that carry water to the generators. The grass keeps down interior temperatures and settles dust at the windy bottom of the canyon.

FIRST OVERLEAF

Only boats can explore the myriad, mazelike canyons of Lake Powell today. Before the waters rose behind the dam in 1963, flooding one of the natural wonders of the world, only a few people had ever seen Glen Canyon.

SECOND OVERLEAF

Dust blankets Lake Powell's Lone Rock Beach on Memorial Day as the regular late evening wind storm kicks up the sands along the beach. Despite the dust, thousands of campers come from California to boat and party on the longest man-made lake in the United States.

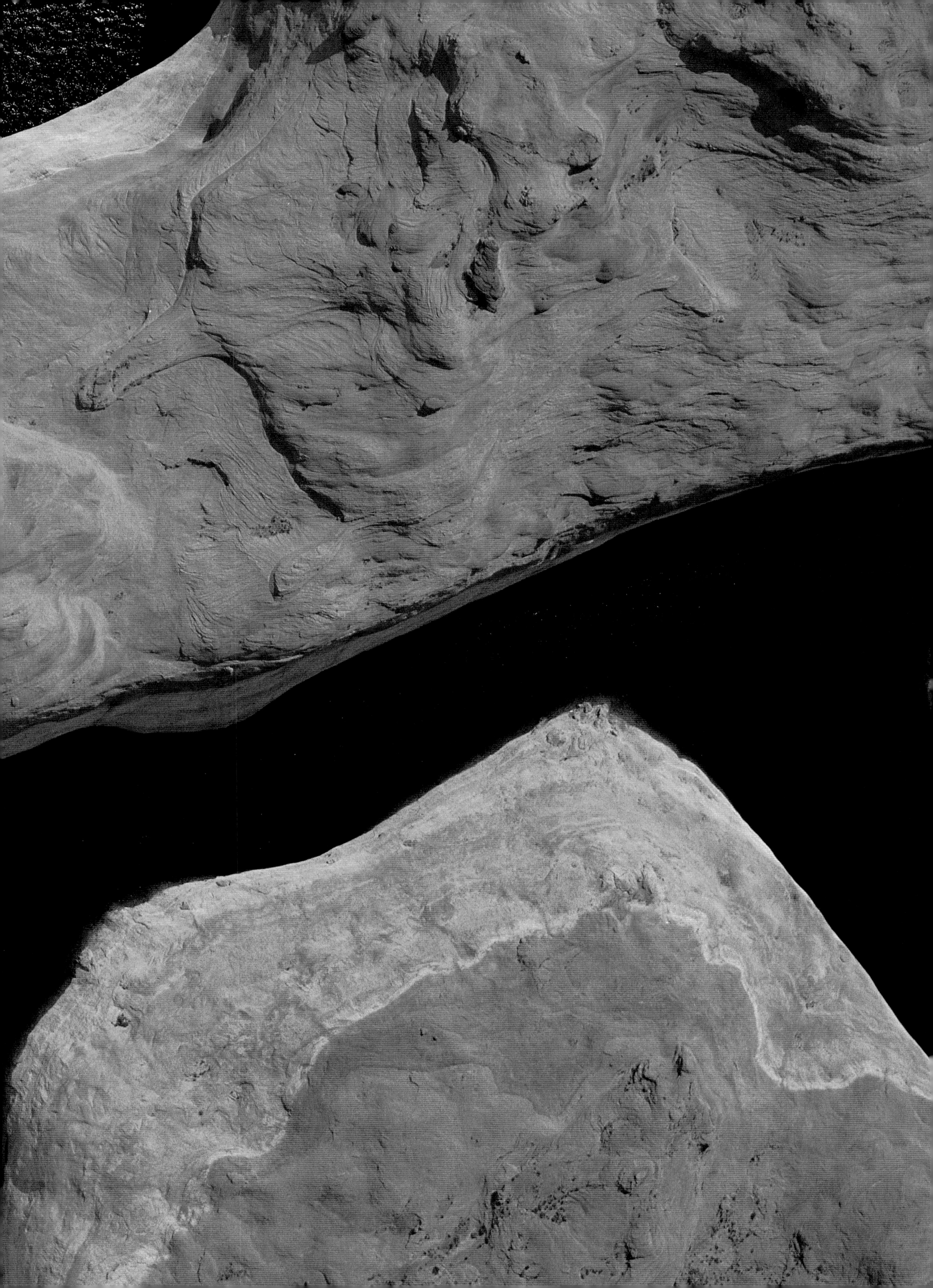

Komfort

RIGHT

Like moths to the flame, the high cliffs of The Cove on Lake Powell attract daredevils. Leaping more than 60 feet , and aiming at a rock-free deep spot, jumpers are actually making their takeoff from the normal shoreline. Years of drought have lowered the huge lake by more than 66 feet, leaving the water-bleached rocks at The Cove looking like a giant bathtub ring.

FAR RIGHT

After a blistering side trip up Saddle Canyon in mid summer, visitors are only too glad to take a shower in the narrow, moss-lined pool. Such pristine sites are saved from crowd damage only by their inaccessibility. While nearly 20,000 people make the trip through the canyon every year, fewer make the hikes into the side canyons.

FIRST OVERLEAF

The afterglow of sunset reveals the moon rising over Lake Powell and 10,388-foot high Navajo Mountain on the Navajo Reservation.

SECOND OVERLEAF

The Fourth of July on Lake Powell, and visitors are having fun. Swimmers plummet from the roofs of luxurious houseboats into the deep green water. In the distance, the mighty Glen Canyon Dam is a thin ribbon of concrete beneath a graceful bridge.

For teenagers from Page, Arizona summer stretches endlessly like the ceaseless shores of Lake Powell. While the very existence of Lake Powell is controversial to many conservationists, it is little more than a fact of life to youth of this generation, a part of the Western landscape for their entire lives.

All-terrain vehicles ply the sands of Lone Rock Beach on a summer evening. As the lake water has fallen with the drought, the beach has grown to more than a quarter of a mile wide. The fresh sands tempt many to drive down to the water where they often get stuck. Locals with four-wheel-drive pickups prowl the beach making money pulling the unwary out.

Although thousands of people can clog Lone Rock Beach for major holidays, Lake Powell's 1,960 miles of shoreline and more than 90 major canyons assure visitors of finding solitude if that is what they are after. But Lake Powell's vastness comes at a steep price. Two million-acre feet of water are lost to seepage and evaporation at Lake Powell and the other reservoirs in the Colorado River system each year.

OVERLEAF

After 33 years in Arizona, Charles Mustrow was ready for his first look at Lake Powell. He found the new parking lot on the Utah side to his liking, set up camp with his RV and folding chairs, and proceeded to count the speedboats. The National Park Service started building more facilities on the Utah side after Utah complained that Arizona was getting all the tax revenues.

ABOVE
Evening brings a golden glow to the luxury boats clustered around the Wahweap Marina on Lake Powell, a visible sign of the wealth tourists bring to Arizona. In the background the Navajo Generating Station, in part owned by the Bureau of Reclamation, sits on Navajo land and burns Navajo coal. It also pollutes the air over the Navajo Reservation, Lake Powell and the Grand Canyon.

RIGHT
Navajo Linda Argo lives with her two children in a tarpaper hogan just a mile east of the Navajo Generating Station. They live without electricity and must carry their water in barrels from a trading post several miles away. Much of the electricity from the plant goes to pump water from Lake Havasu, hundreds of miles to the west, to Phoenix.

OVERLEAF
Irrigated fields crowd badlands at the Navajo Irrigation Project near Farmington, New Mexico. The Navajo Agricultural Products Industry uses water brought by canal from the Navajo Reservoir on the San Juan River to grow a number of crops, including wheat, corn, pumpkins, beans and decorative squash. Water towers provide pressure for center-pivot irrigation.

Left
Irrigation creates emerald green fields of alfalfa at the base of Castle Rock, north of Moab, Utah. Famed for its rugged red rock landscape this area is a favorite shooting location of movie producers and advertising executives from around the world. The Moab Film Commission aggressively markets the landscape for everything from pickup truck ads to European cigarette commercials.

Above
Halloween means pumpkin harvest on the Navajo Reservation and lots of short-term minimum-wage work for tribal members.

Under a misty canopy that keeps the New Mexico desert sun at bay, Navajo Matthew Endischee harvests Shiitake mushrooms. Part of the Navajo Tribe's effort to make every possible economic use of their share of Colorado River Basin water (that they get from the San Juan river), the mushroom crop goes to gourmet restaurants in San Francisco, Los Angeles and Washington, D.C.

Left
Grand Canyon visitors become steeped in its geologic history as their boats take them deeper and deeper into the ancient rock strata. Resting gently in a sheer rock cove out of the ferocious rapids and the searing heat at the bottom of the Grand Canyon, a dory is the image of tranquil grace, at home in the canyon.

Far Left
Foaming and seething, runoff from a distant thunderstorm boils across the red earth in Monument Valley.

First Overleaf
Below Lee's Ferry the Colorado River cuts into the Paria Plateau forming Marble Canyon. Between here and Lake Mead the river drops 1,900 feet and half of that takes place in the 160 rapids that make the Grand Canyon a favorite destination of river runners.

Second Overleaf
Pulling hard through Grand Canyon rapids, conservationist Martin Litton works to keep his dory upright and off the rocks. At 75, Litton is considered by many to be the grand old man of the Grand Canyon. Nearly half of his life has been devoted to the fight to keep the canyon wild.

In 1540, Captain Garcia Lopez de Cardenas, with help from Hopi guides, became the first European explorer to see the Grand Canyon, but he was unable to reach the Colorado River far below. Two centuries passed before European explorers came again.

Seven hundred feet above the canyon floor, Nankoweep Granary, an Anasazi ruin, offers a stunning view of the Colorado River at sunrise. Archaeologists debate whether these structures were for food storage (if so why include windows?) or housing in a remote and defensible position. The Anasazi seem to have used the Grand Canyon only sporadically for homes, but maintained trails across it.

OVERLEAF
Red Wall Cavern has been a favorite of Grand Canyon voyagers since the days of John Wesley Powell. To celebrate his love for the place, dory boatman Pete Gross plays a bit of Beethoven's Ode to Joy on his recorder.

ABOVE

Tides surging through the Grand Canyon erode beaches. Water levels fluctuate wildly as Glen Canyon Dam releases water when cities in the Southwest demand electricity. On a hot day when Phoenix turns on its air conditioners the Colorado River rises – a lot. At Nankoweep the river rises five feet overnight (to levels corresponding to streaks of light on the beach) and lifts the dories with it. Many beaches are much smaller today than only a few years ago.

RIGHT

Farther down river, the dam-initiated tides take more than 12 hours to arrive, and sunrise in the Canyon finds the river at low water. The beaches, alternately flooded and drained every day, show the effects of the erosion.

64

Rafters spending the night on a beach at South Canyon awake to find the river very low ("Must have been a cool day in Phoenix yesterday," a boatman complains) and their rafts are stranded. Pushing the giant rafts to the water requires everyone's help.

OVERLEAF
Rafters in the Grand Canyon discover why they call the Colorado River wild. Most rapids in the Canyon occur where side canyons wash sediment and boulders into the river, creating a gentle, dammed pool above and a rampaging ride below. Canyon rapids are rated from 1 to 10 (with 10 being the wildest) and can vary widely in difficulty depending on the water volume.

Visitors to the Grand Canyon, isolated from the outside world, adapt to Canyon time where change is measured in geologic ages as revealed by the surrounding rocks. While her passengers explored Red Wall Cavern, a dory boatwoman finds time to read.

OARS

CHAPTER THREE

THE LOWER REACHES

When Major John Wesley Powell emerged from the Grand Canyon in 1869, after three months and 200 rapids, he met Mormon colonists who were making the desert bloom. They gave him watermelon, bread and cheese, produced on land that received only four inches of rain a year. The spot is now under Lake Mead, which abruptly ends the river's wild streak. Powell later foretold the opportunities and limits of western water: "All the waters of all the arid lands will eventually be taken from their natural channels," he wrote. "There is not sufficient water to supply the land."

The Mormons believed that irrigation fulfilled the prophecy of Isaiah, that when Christ returned "the desert shall rejoice, and blossom as the rose...for in the wilderness shall waters break out, and streams in the desert. And the parched ground shall become a pool, and the thirsty land springs of water."

Like a miracle the river plugged by Hoover Dam, on the border of Nevada and Arizona, achieved that promise. The "grande dam" reined in the Colorado, fostering the richest irrigation project in the world, and powered the Sunbelt. Hoover Dam, built in 1935, created the template for future western dams—the "cash register" power plants that pay back not only the cost of the dàm but finance other dams and irrigation projects as well. Hoover was a "winner in all aspects," said Julian Rhinehart of the Bureau of Reclamation.

Lake Mead holds yet another two years of the river's flow. But because it lies in a desert climate, it loses five feet of water a year just to evaporation. Mead, Powell and the other, smaller, reservoirs on the river lose two million acre-feet of precious water each year into the air. Like Powell, Mead also stops the tons of mountain and canyon silt that flow to the sea. The Bureau of Reclamation estimated that without Glen Canyon dam upstream, Lake Mead would be half-full of silt by now. In stopping silt, Glen Canyon also eliminates nutrients like phosphorus flowing into Mead. Algae depend on phosphorus and algae is food for plankton, the tiny building block of the food chain. As a result, Mead's famous bass fishery is dying. One scientist said he began catching bass that looked like eels, long and skinny. Some fish groups have proposed fertilizing Lake Mead to boost the fishing economy.

By controlling floods, the dam made possible riverbank developments like the "Colorado Belle," a giant casino built in the shape of a sternwheeler in Laughlin, Nevada, 50 miles below Hoover. Neon lights light up the sky with signs flashing "Slots" and "Chicken buffet $2.99." When Don Laughlin bought an old riverside bar in 1966 there was nothing but sand and little need for water. Today there are a dozen casinos with 15,000 rooms in a boomtown race for both money and water. Ferries ply the 100-yard-wide river night and day bringing gamblers from parking lots on the Arizona side. But while the river is wet and inviting, Laughlin can use only about eight hours of the river's flow for the entire year. "They see the Colorado outside and say, 'What do you mean, you're running out of water?' " said Ted Finneran, a casino employee who chaired the town board. Yet 10,000 acre-feet — Laughlin's share — "can only be spread so far," said Jim Ley, with the Clark County Department of Comprehensive Planning. Laughlin uses 4,000 acre-feet but developments have been approved that will take more than that. Thus, Laughlin is the first Sunbelt community to feel the limitations of the river. "Laughlin will be the first community to face growth limits in the mid 1990s," said Ley. "Nothing has ever restricted enterprise before. They don't cotton to that kind of talk. In reality, it's fact."

Laughlin's trouble is just the tip of the iceberg for Nevada. When the waters of the Colorado were divided by the interstate compact in 1922, Nevada had no irrigated agriculture, and Las Vegas was a town of 5,000. As a result, Nevada got only 2 percent of the water. That share is virtually all used. With 6,000 new residents per month, Las Vegas is looking for water — anywhere. Officials have even considered shipping in water from Alaska. The Las Vegas Valley Water District looked to the interior of Nevada, filing water claims on groundwater beneath 20,000 square miles of Nye, Lincoln and White Pine counties. A 1,200-mile pipeline could bring water from wells drilled all over the desert, a multi-billion dollar project three times bigger than the Central Arizona Project, farther down the river. Opposition sprang up immediately from as far away as Death Valley, which claimed that much drilling would dry up springs in the park,

Imperial Valley vegetable grower Jim Storm displays his "River Ranch" belt buckle. Water makes wealth in the arid valley.

possibly harming the endangered Devil's Hole pupfish. Another option open to Las Vegas is leasing or buying unused water from Wyoming or Colorado.

Nearly 150 miles south of Laughlin, on Lake Havasu, the sound of slurping straws can be heard. On opposite sides of the lake two huge pumping stations suck water for distant civilizations. Not far from the famous London Bridge, the Central Arizona Project carries river water 335 miles eastward to Phoenix and, soon, to Tucson. On the other shore is the Colorado River Aqueduct, emerging from a pump house capable of sucking up to one billion gallons of water a day for Southern California. The Metropolitan Water District of Southern California, commonly called the Met, which runs the intake pipes, was formed in 1928 during the Depression. Voters agreed to spend $220 million to build the aqueduct. "People voted not for water, but for jobs," said an engineer on the project. The first water flowed in 1941. Water is lifted 1,617 feet, largely from power generated at Parker Dam that backs up Lake Havasu. Each of the eight pumps could fill a city swimming pool in eight seconds.

Until 1990 Californians took pretty much what they needed — over one million acre feet a year, twice the Met's right to the river and about 60 percent of Southern California's demand. The reason: none of the other states on the river was using its share, so there was extra water in the river. With the Central Arizona Project nearing completion on the other side, however, Arizona began to take more. In 1990, when dry California asked for its usual allotment, the Bureau of Reclamation, with its hand on the Hoover spigot, said no. It was an historic announcement. The lower basin had, for the first time, used up its full share of the Colorado River.

Given these limits it was strange to travel west along the Colorado River Aqueduct and find its waters spread in chevron-shaped shallow ponds near Palm Springs, soaking into the ground! Like all desert mirages, the sight wasn't what it

Paralleling the Mexican border, the All-American Canal carries one-fifth of the Colorado River through sand dunes west of Yuma, Arizona.

Computers take control of the water in the control room of the Central Arizona Project in Phoenix. Lights blinking at various pumping stations on the wall-sized map trace the canal's route across the desert. From here, operators can control any gate or pump on the 335-mile aqueduct between Lake Havasu and Tucson.

appeared: The water was recharging the desert's huge underground aquifer. About 55 percent of the water used in the Palm Springs area comes from the river. Without the Colorado River, the Coachella aquifer would drop by three feet a year, according to the Coachella Valley Water District. Much of the water poured into the sand is being "banked" by the Metropolitan Water District. When it needs water, the Met withdraws its savings by taking water in an exchange with the Coachella district.

Because of the rich aquifer, wealth beyond all imagining has come to Palm Springs. On Country Club Drive I passed sumptuous homes behind walls and guardhouses, developments with names like "The Lakes" and "Desert Falls." At the 400-acre Marriott Desert Springs resort, swans glide on 20 acres of artificial lakes. The lobby, a vast atrium, is festooned with palms and vines and two waterfalls cascading into a 30-foot-deep indoor lagoon, complete with a dock and boat that ferries guests to hotel restaurants. The lake holds 20 million gallons, pumped from the aquifer. From the lakes that surround the resort, water is pumped onto a golf course that uses one-and-a-half million gallons of water a day in summer.

"One thing I've noticed in living here 30 years is the humidity level has increased because of all the watering of golf courses," said Lee Jesson, an engineering supervisor who took me on a tour in a golf cart. We stopped by the 18th hole with a 10-foot waterfall. There are so many water hazards that skin divers recover golf balls by the hundreds and resell them to the resort.

"As best we can tell we haven't lowered the water table," said Don Murray, the resort's director of engineering. The resort's second golf course is using reclaimed sewage water, one of eight to do so in the valley of 90 golf courses. The hotel also installed low-flow showers and toilets in its 900 rooms. But there is little push for desert landscaping. "Some of our image is the Florida-type of flowers and grasses," said Murray. "The only real problem we have is the politics in the other part of California. Los Angeles and Santa Barbara look at us and say we're doing nothing."

Near Riverside, California, the water from the Colorado is mixed with water coming from northern California and is divvied up among 26 cities and San Diego County. Metropolitan Water supplies almost half of the three million acre-feet used by 13 million people each year, but Southern California is growing at the rate of 300,000 people a year — the equivalent of adding a city the size of Portland every year.

In Los Angeles I found serious attempts to conserve water as the California drought entered its fourth year. Squads from the city's Drought Busters enforced new ordinances against washing sidewalks, serving unsolicited water in restaurants and watering lawns during the day. They also gave out hose spigots and installed low-flow shower heads. I spent one morning cruising the streets with Drought Buster Tony Marufo who braked to a halt at the first sign of a damp spot on a hot sidewalk. "Some people say I can smell water," he said, grinning. He was wearing his "uniform," a T-shirt with a Drought-Buster logo, modeled after the "Ghost Buster" sign. "Fifty percent of all leaks are on the street, so you follow the water trail and see what's at the other end." He usually gives people three days to fix a leak.

The water crisis in Los Angeles was real. A pro-environment court ruling had restricted the city's use of its Mono Lake and Owens Valley water supply, so the city turned to the Met, where it has an entitlement of up to 25 percent of the Met's water. As 1990 began, Los Angeles got 60 percent of its water from the Met compared to 10 percent in a normal year. In a city long known for its profligate water use, Marufo and his 25 colleagues wrote 8,862 citations from May to October. By late summer Los Angeles had reduced use by more than 10 percent, a goal set by the Met, which imposed first-ever conservation measures on its 27 members.

"This is going to be a way of life," said Jim Derry, director of customer service for the Los Angeles Water and Power Authority. "We've got 300,000 people coming to Southern California every year. They're going to need water, too."

But the Met's Tim Quinn told me that conservation is a limited tool, that per capita use is up in most western cities. Newer houses actually push water use up — all have automatic dishwashers — and higher income families use more water.

"We're trying to find ways to flatten those numbers out, but lowering them may be impossible. It would cross the line of fundamental changes in life-style — no green yards, for example."

And that, I learned, was a line that no one in the California water establishment wanted to cross. Controlling growth, they said, is not a water agency's job; finding more water is. Officials fear running short of water by the year 2000 if serious drought conditions continue. As Arizona takes its share of the Colorado, Southern California will be forced to look north, sending reverberations all the way to Marin County, across the Golden Gate Bridge, north of San Francisco. The Sacramento River and the unfinished State Water Project "soon must replace the Colorado River as the backbone of the Met's water supply system," said Duane Georgensen, assistant general manager for the Met.

Many northern Californians object, saying southern Californians are stealing their water. They say conservation and reclamation of waste water could handle the increased demand in the south. If more water is taken from the Sacramento River, it will reduce fresh water flows into San Francisco Bay, already dying because of inland water diversions for agriculture. The fishery in the bay has dropped by more than 50 percent, with the loss of 2,000 fishing boats.

"They're losing the Colorado River, they're losing Mono Lake water. And what that means is they are looking for any developed water" said Steve Roberts, a fishing spokesman. "Driveways are still wet in California and still they say they need more water."

The drought also has raised anew such oddball possibilities as sea-going tankers bringing fresh water from the Pacific Northwest, ships hauling icebergs from the Arctic, and building a pipeline to tap Washington's Columbia River. Desalination plants are under construction to turn Pacific Ocean water into fresh water. But desalinization takes tremendous energy. Another prospect is water marketing — trading water like a commodity, a relatively new concept in California. In its first deal the Met

Partyers tie their boats together in Lake Havasu's Copper Canyon. After climbing the cliffs, revelers plunge to the water 50 feet below.

agreed to finance the lining of irrigation canals and the upgrading of Imperial Irrigation District plumbing at a cost of $223 million. The deal will save 100,000 acre feet per year from seeping into the ground, thus increasing water available to the Met. The agency might also pay farmers not to grow crops in dry years.

Agriculture uses 80 to 90 percent of the Colorado. "Clearly the greatest conservation source is agriculture," said Quinn. "They get 85 percent of the water and they have not been under the political pressure to conserve." The Met also has discussed leasing water from Indian Tribes and upstream states. "The phone rings three times a week from landowners in other states" wanting to sell their water, said Quinn. Wyoming might use its water after all — by selling it to California. But Wyoming laws would have to change for that.

Everywhere I went along the river a new breeze was blowing on water policy. But at the end of the pipeline, where I expected a hurricane force, it was a whisper. One of the fastest growing communities in Southern California is Chula Vista, outside San Diego — virtually dependent on the Colorado River. EastLake, a development being built there on barren land a few miles from the Mexican border, is an example of how water leverages growth. It is literally the last housing development on the river pipeline. A few dollars invested in water — EastLake's water cost of $525 an acre-foot (a western family of five uses one acre-foot of water a year) helps turn worthless ground into a thriving community, with people, schools and traffic."This is going to be our downtown here," said developer Bob Snyder, pointing to a gully of sand. Over another hill we came to a reservoir called EastLake. The company's "Hills and Shores" subdivision, with 1,834 homes ranging from $90,000 to $325,000, sold out in four years. Peer pressure keeps most homes surrounded by bluegrass. Snyder said he could not force xeriscaping, the use of water-miser desert plants. "Maybe in five years I can lead them to it," he said as he drove me by a beach club, where palms and grass surround a sandy beach. "This is the kind of amenity that brings people to EastLake. We've spoiled ourselves, no doubt about it." In a new group of homes built around a golf course, Snyder hopes to reduce the average per capita consumption of water of 110 gallons a day by 30 percent.

Snyder, an athletic third-generation California builder, does not think his company is causing growth, but is responding to it. Indeed, half the growth is from new births. "If we had an unlimited supply of water tomorrow, growth would still be a problem. People want homes. We know how to do that. If we're good at it, people will buy our product." When a limited-growth proposal went on the ballot, Snyder's company opposed it. The voters rejected it. "They went too far. Most people agree that some growth needs to occur. Some level of growth is important, to enhance the quality of life. The development industry is stereotyped as being for rampant growth. I think that's a bum rap. Conservation alone won't sustain a reasonable growth rate."

An avid white-water canoer, Snyder admits to mixed feelings about draining Northern California's rivers for his development.

"I don't want to degrade anybody's river. There is enough water in California for whatever we want to do. We just have to decide what we want to do. We have to make tradeoffs. Maybe we have to raise the price of water to make people save and put the money into desalinization. We get calls all the time from Colorado, about selling water to us." When I asked Snyder if he expected to ever reach a limit, he said no. "The professionals in the water supply business articulate it as a plumbing problem, not a water supply problem. Once we get through a drought for the near term, there will be water available for Southern California — that's what I hear from the water experts."

By the end of summer 1990 it had rained in Arizona, demand on the Central Arizona Project declined and California was able to take what it wanted from the Colorado River. Without drought hanging over its head, Southern California is not likely to conserve its way into the 21st Century.

Saguaro cactuses stand guard in Arizona along the ribbon of water called the Central Arizona Project. This $3.5 billion Bureau of Reclamation network is considered by environmentalists the ultimate in desert folly. In order to pump 1.5 million acre-feet of water to Tucson per year, and subsidize its cost so farmers can irrigate economically, the bureau helped build a coal power plant near Page, Arizona that pollutes

The fountain at Fountain Hills, east of Phoenix, spews water 630 feet into the air, higher than any other fountain in the world.

the air over the Grand Canyon and Navajo Indian Reservation.

South of Phoenix, on the tiny Ak-Chin Indian Reservation, the same Central Arizona Project is a godsend, however. Water from the Colorado has turned a dirt-poor community into a prideful, self-sufficient farming community with almost no unemployment and no welfare. "We're using it to farm cotton, small grains, alfalfa — even fish — and 95 percent of our people work on the farm," said Leona Kakar, a vibrant woman whose family led the way to this Indian-country success. She spoke to me in the shade of a tamarisk tree after ceremonial basket dances to celebrate ground-breaking for a tribal museum. A mile away, workers were harvesting cotton. "I've given 26 years of my life for this fight," she said with a shake of her graying, curly hair. "It's made a world of difference."

Until the 1960s a few hundred Ak-Chin subsisted in the Sonoran Desert around shallow wells. But pumping around Phoenix lowered the water table hundreds of feet, making water for farming too expensive. Citing the 1908 Supreme Court Winters doctrine, which reserves enough water for Indians to irrigate their land, the tribe sued the federal government. The first Congressionally-awarded water from the Central Arizona Project arrived at the Ak-Chin farms in 1987. Within two years a 38 percent unemployment rate had dropped to 4 percent, and the tribal farm had tripled its acreage.

Encouraged by the Ak-Chin success, other Arizona tribes, which have been without adequate water for a century, are following suit. Their claims total more than all the water in Arizona, which arguably could make them the American Arabs of water. The huge Navajo reservation, for example, could claim the entire flow of the Colorado, based on the Winters doctrine. "A cloud of uncertainty hangs over our water management programs," said Larry Linser, deputy director of Arizona's Department of Water Resources.

But a water right without water is useless, and several tribes have compromised in order to get canals built and water delivered. The Central Arizona Project carries Colorado River water for 10 tribes. "We are doing what we did in A.D. 200, just a little more modern," said Leona Kakar. "Water is our lifeline, our blood."

Below Lake Havasu the Colorado runs like a sluice, wide and sluggish, the banks riprapped in places and lined with tamarisk and occasional rows of cottages. Near Yuma, Arizona, it backs up behind the Imperial Dam, which takes more than 20 percent of the water – the single biggest chunk of the river – and pushes it through the chimerical All American Canal about 80 miles west to California's Imperial Valley. It is the start of the river as a plumbing system. Most of the water now runs in canals, "operated with hydraulic efficiency" through generators and pumps.

Driving west I watched dunebuggies race over the sand dunes sliced by the canal, where 70,000 acre-feet of water soaks into the sand each year. The Imperial Irrigation District, at the other end, has a very senior and generous allowance of the river's water but even here the 700 farmers on 464,000 acres are feeling the pinch of public pressure. California cities can't understand how so much water just disappears into the ground. At their urging Congress has authorized the Bureau of Reclamation to stop the leakage on the canal. Lining it is a likely solution.

The cities also can't understand how nearly one million acre-feet of water — a third of the valley's water rights — runs off and under irrigated fields and into the briny Salton Sea, which lies 200 feet below sea level. The state of California has found that the district, at the far end of the canal, wastes about 15 percent of its water and has required the irrigation district to conserve 100,000 acre-feet per year. After 90 years, farmers who grow nearly $1 billion in produce are beginning to worry.

"We do live in a democracy," said Larry Cox, a 32-year-old cotton, alfalfa and vegetable producer I found weighing onion seed. "If you've got 16 million voters up there in L.A., who's to say they can't change the laws? I think it's a definite possibility. We use a lot of water down here. We have irrigation water running someplace on the ranch probably 360 days a year. With this system we need the surface runoff to prevent the soil from salting up."

The land is so salty and the river water is so saline (averaging more than 700 parts per million) at the end of its journey that extra water is poured through the soil to flush salt away from crop roots. Each field gets about four feet of water. Pipes buried four- to eight- feet below the surface carry away the excess, salty water.

In a corner of the field behind him, trucks pouring concrete

The Los Angles skyline shimmers in the reflecting pools surrounding the Metropolitan Water District of Southern California.

were paving a lateral canal, part of the Met exchange. Cox's father, Don, who sits on the irrigation district board, lined all his field canals to reduce salt leaching. He also graded fields more precisely so that plants at the end of the row get as much water as those at the beginning. Don Cox is pushing the district to reduce runoff water, to cooperate in water marketing, and perhaps to take land out of production to slake the thirst of Los Angeles, but he does not speak for the entire district.

With the precision of a druggist, Cox measured the onion seeds to get 25 plants in one foot of the 40-inch-wide bed, 1,680-feet long. Punching a calculator after two test runs, with the planter's efficiency and the seed's germination rate, he reduced the seed pile by an ounce or two. The onion seeds would be watered by portable irrigation pipes until they germinated in six days. Then water from the Colorado River would run down the laser-perfect rows until the onions are harvested nine months later in mid-May. They will be dehydrated and used on McDonald's hamburgers.

"If we have a perfect stand, I can make 25 tons per acre, at $75 per ton," said Cox. He has 1,000 acre planted in onion and hundreds more in lettuce. Despite its reputation as a vegetable plot, the Imperial Valley's biggest crop is beef ($222 million annually) followed by alfalfa ($166 million).Lettuce grosses $126 million.

"Produce is risky," said Cox. "Last year there were guys who lost $1,000 an acre because of poor crops and the markets. If we can make $200 an acre net, I'm ecstatic. It's not a utopia down here by any means. I spend a lot of my time thinking that we're going to go broke."

Cox said he resisted more efficient methods of irrigation because of the cost — $300 an acre and up — until he was forced to try it on a tomato field plagued by water-borne soil disease. "It was fairly easy, and the yield went up substantially."

LEFT

Completed in 1935, the 726-foot-high dam was an engineering marvel that lifted the spirit of America during the depths of the Depression. It promised cheap and abundant power and water for the foreseeable future of the Southwest. Lake Mead, spreading across the purple desert, can store two years' flow of the Colorado River.

ABOVE

Visitors touring Hoover Dam are invariably awed by the soaring vistas and the incomprehensible numbers: 4,360,000 cubic yards of cement weighing 6,600,000 tons; water pressure of 45,000 pounds per square foot at the base of the dam; a reservoir that will hold 28,537,000 acre-feet of water; and 800,000 acre-feet of water lost to evaporation every year.

Far Right

The dam's 17 generators are housed in two cavernous rooms, each 650 feet long. Water from 500 feet above comes down to the turbines through 30-foot-diameter penstocks. Though the dam was completed in 1935 the final generator was not installed until 1961 in response to growing demand for power. The Metropolitan Water District of Southern California is the largest single buyer of power, much of which goes to pump water to Southern California.

Right

Hoover Dam's gigantic generators are being retrofitted to increase their generating capacity from roughly 1,300 megawatts to 2,100 megawatts. This huge 23-foot rotor was pulled from the generator housing to be rewound with modern windings.

First Overleaf

Like a desert serpent, the $3.5 billion Central Arizona Project canal twists through cactus-strewn hills. Carrying half a trillion gallons of a water a year, the Bureau of Reclamation project will allow continued large scale growth in arid central Arizona.

Second Overleaf

New houses march across the desert in Palm Springs. Water from the Colorado River Aqueduct not needed immediately is pumped into settling ponds west of the city where it soaks into the underground aquifer and is stored for later use.

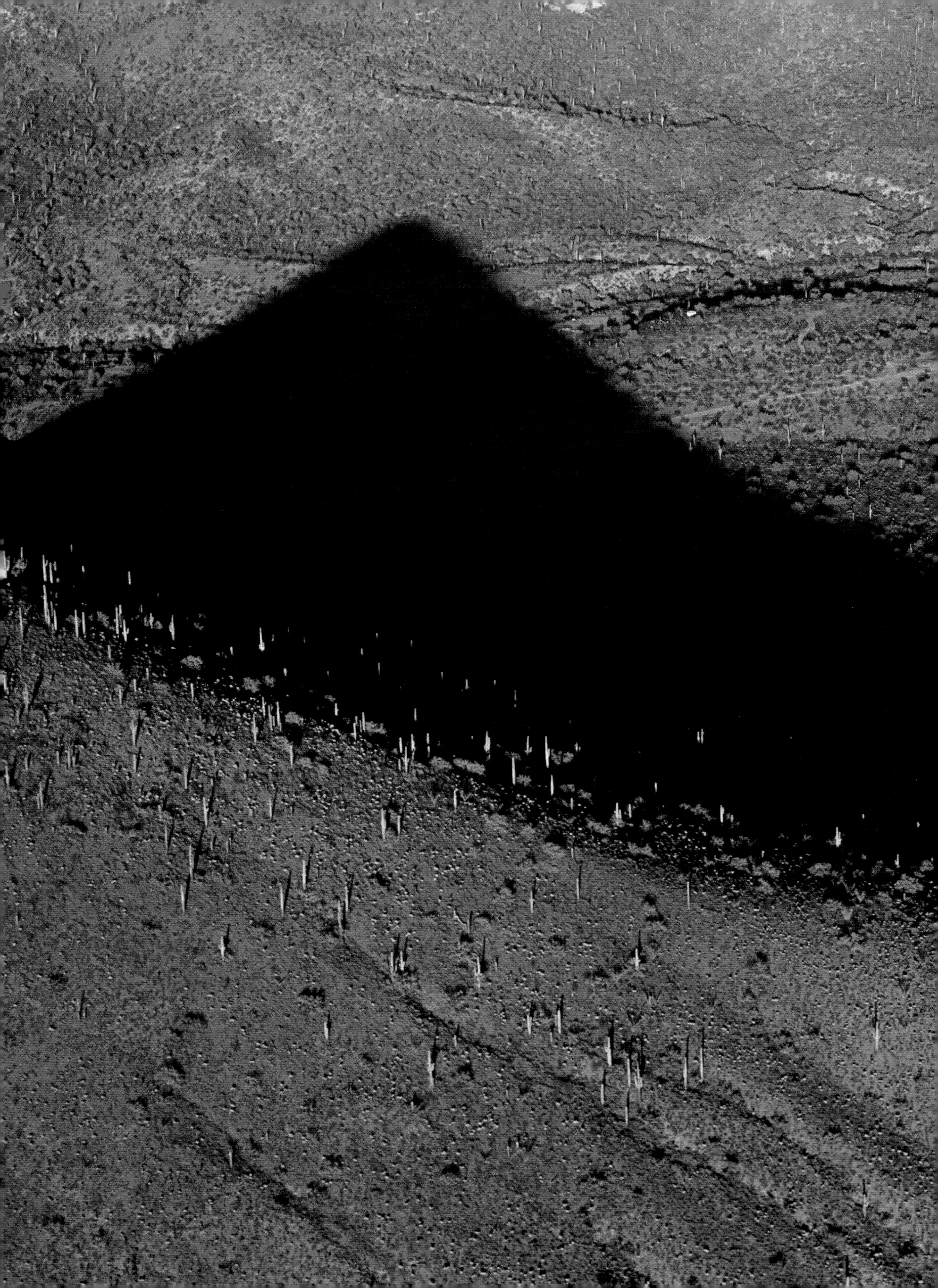

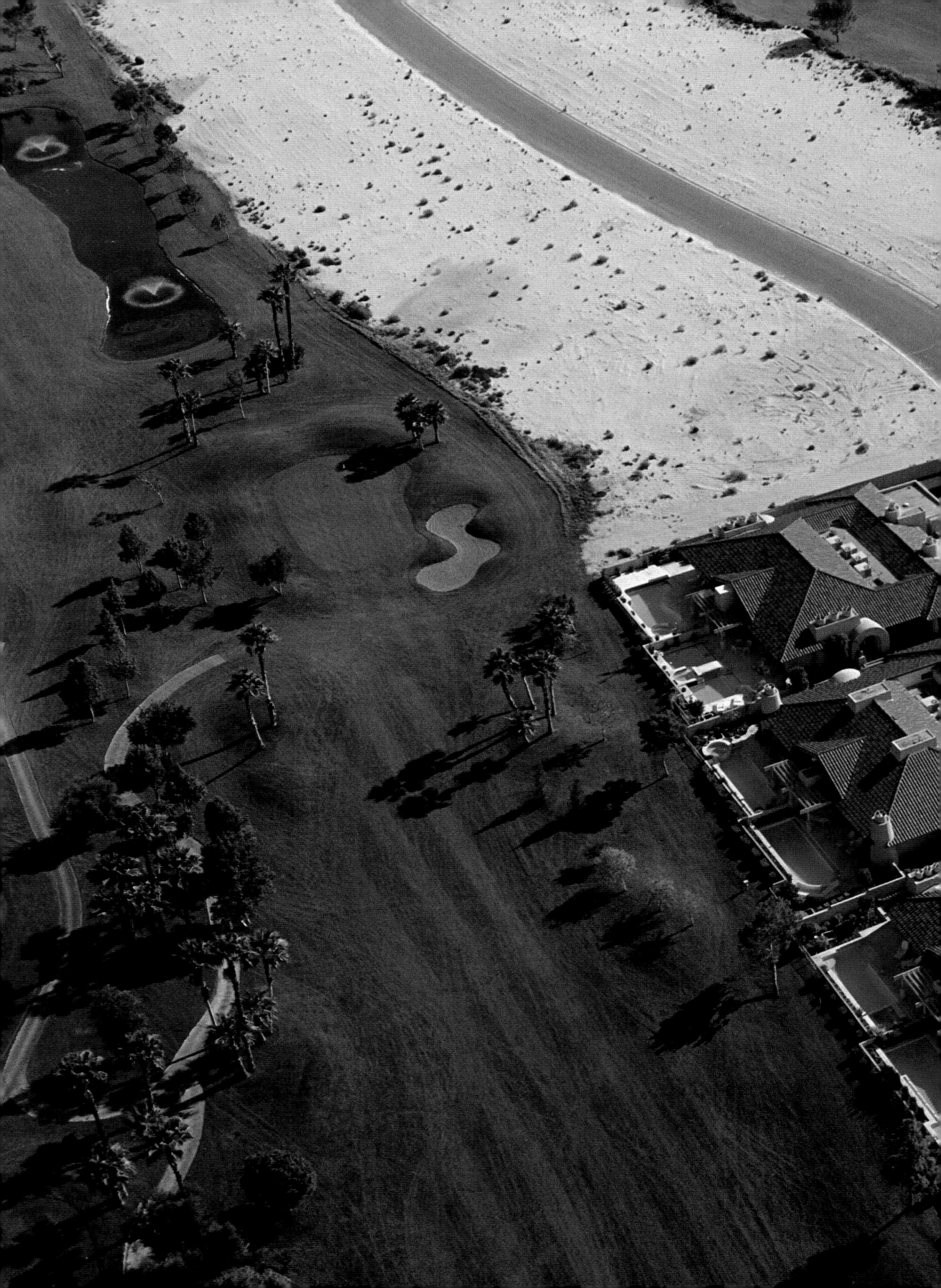

On one side of this Palm Springs street are desert dunes; on the other side, lush lawns. The difference is water.

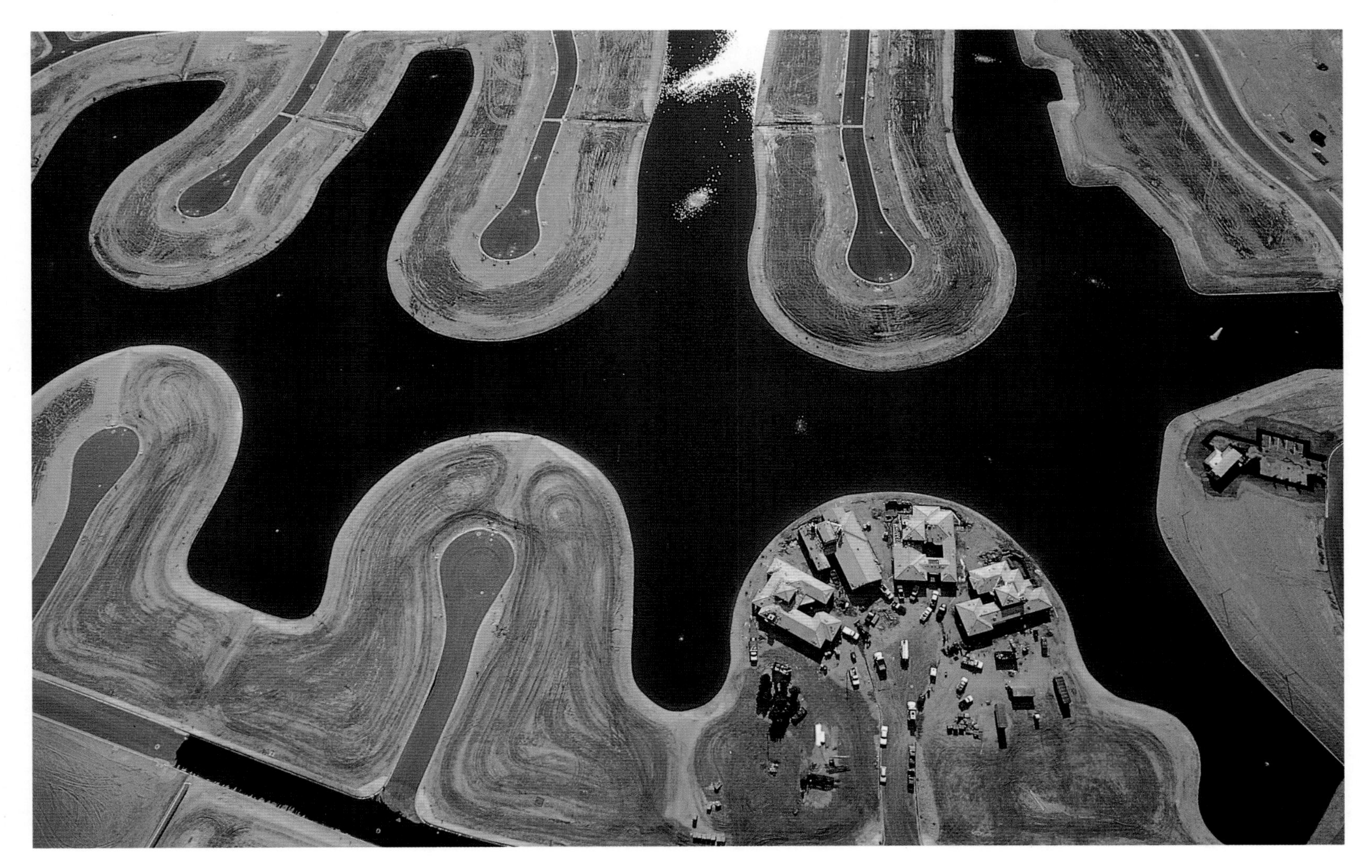

This mini-lake in the Lake La Quinta development near Palm Springs will attract home buyers, each of whom will get a waterfront lot. These lakes are cement-lined to prevent water seeping into the sands. Colorado River water reaches Palm Springs via the Coachela Canal and the Colorado River Aqueduct.

OVERLEAF
Waterfalls cascade into an indoor lake at the Marriott Desert Springs Hotel. A prime example of the southwestern art of water sculpture, the hotel lobby is a lush retreat from the surrounding sands. Tour boats dock in the lobby and take guests out through sliding doors, past resident flamingos, to a nearby restaurant or on a tour of condos for sale.

Shimmering lakes surround the greens at the Marriott Desert Springs resort near Palm Springs. It takes 1.5 million gallons of water a day to keep the fairways lush and green. Just across the highway are sand dunes.

THE MIRAGE

LEFT

Gas pipes concealed underwater create the illusion of a flaming volcano at the Mirage in Las Vegas. But while casinos create incredible water sculpture using recycled water to lure gamblers, the City of Las Vegas is on the verge of running out of water. With its slim slice of the Colorado River already used up, the city is seeking to drill for water in rural Nevada to keep its phenomenal growth on track.

ABOVE

Captain Claude stands ready to take another group on a tour of the man-made channel that now carries water under the London Bridge, which was moved from England to Lake Havasu. The bridge now takes visitors to vacation homes on an artificial island.

OVERLEAF

Like water beetles scurrying across the surface, ferries take gamblers across the Colorado to Laughlin, Nevada. With a limited water allotment in Nevada, gambling developer Don Laughlin put his first casino, the Riverside, on the Nevada side of the river and motels, houses and parking lots in Bullhead City on the Arizona side.

PIONEER PIONEER
HOTEL
& GAMBLING HALL
RAMADA EXPRESS

COLORADO BELLE

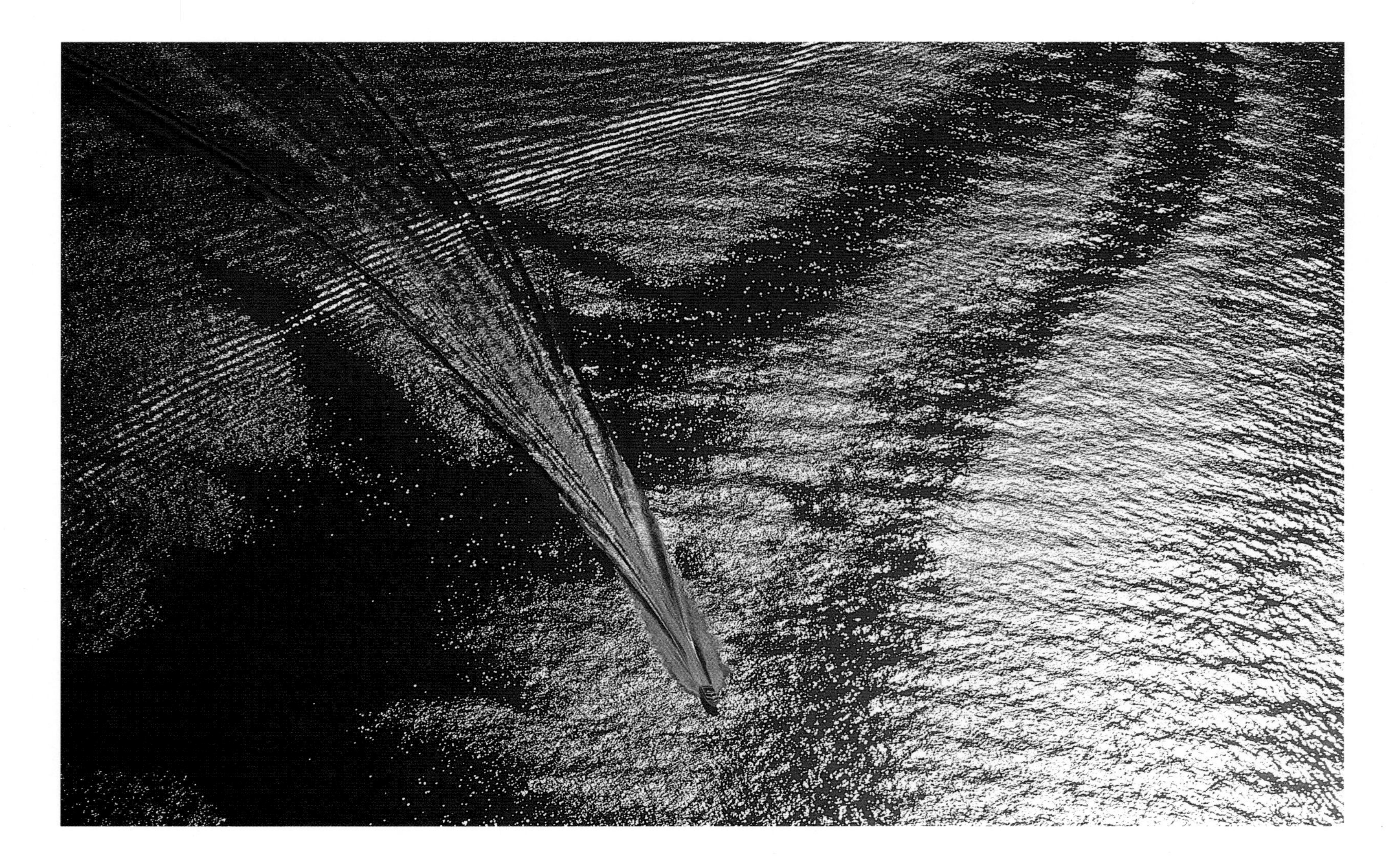

Early visitors to the Colorado River found it uniquely inhospitable. In 1858, Lt. Joseph C. Ives wrote, "Ours has been the first, and will undoubtedly be the last, party...to visit this profitless locality. It seems intended by nature that the Colorado River, along with the greater portion of its lonely and majestic way, shall forever be unvisited and undisturbed."

Convinced he could attract home buyers to the Arizona desert, developer Robert McCullough bought the London Bridge and shipped it across the Atlantic where it was reassembled on a peninsula on Lake Havasu. A canal was then dug beneath the bridge to allow the waters of the Colorado River to flow underneath. He was right. Today, 1.5 million people visit the bridge every year, and Lake Havasu City is booming.

LEFT

The giant pipes at the Whitsett Pumping Station start water from Lake Havasu on its long trek across the desert to Los Angeles. The first facility in the Southern California Metropolitan Water District's Colorado River Aqueduct, the pumping station draws more than 1.2 million acre feet of water from the lake every year, about a third of Southern California's municipal water.

ABOVE

The Central Arizona Project has finally made it possible for Arizona to use its full share of Colorado River water. As a result, California no longer receives the excess water it has come to rely on.

Far Left

Monster speedboats roar past waterfront homes on the Parker Strip below the Parker Dam in California. Canals paralleling the streets allow homeowners with boats access to one of the most popular strips of water on the Colorado River. Holiday boaters pull up to a shoreside bar to watch a wet T-shirt contests and down a few beers.

Left

With all three 1,800-horsepower pumps going, the fountain at Fountain Hills, Arizona can shoot water 630 feet into the air. The brainchild of Robert McCullough, the same man who brought the London Bridge to Lake Havasu, the fountain was meant to be a unique landmark in arid Arizona and to attract house buyers.

RIGHT

Drought Buster George Verdesoto pokes at a trickle of water on a Los Angeles street to see if it is a water main leak or dreaded water waste. Two dozen Drought Busters scour Los Angeles neighborhoods writing citations to anyone wasting precious water. Offenses include watering sidewalks, car-washing without an automatic shutoff hose nozzle, or just watering at the wrong time.

FAR RIGHT

Bringing fresh vegetables to the nation all year round, California's Imperial Valley farmers employ thousands of migrant workers for the laborious harvesting. Today 700 farmers share 2.9 million acre feet of Colorado River water a year, nearly three times as much as the Metropolitan Water District of Southern California gets for nearly 16 million urban users. But from the farmers' point of view, those 16 million people are getting some of that water every time they eat a salad or cook a vegetable.

CYPRESS
CYPRESS
CYPRESS
PLEASE KEEP REFRIGERATED
PLEASE KEEP REFRIGERATED
WESTERN VEGETABLES
34°F REFRIGERATE 1°C

Sprinklers water fields being prepared for another vegetable crop. Sprinklers must be used sparingly because the salty water builds up on the soil surface. They are removed before planting and the crops row irrigated to flush away as much salt as possible. Getting their water claims in nearly 90 years ago, the developers of the Imperial Irrigation District were just following time-honored western water law–"First in time, first in right."

Larry Cox, one of the farmers who shares the bounty of the All American Canal, checks the moisture in a field of onions destined for McDonald's hamburgers. Ironically, Cox finds rain an impediment. He wants water only when his vegetables need it. He can order it at will from the irrigation district offices. Imperial Valley farmers produce $1 billion worth of vegetables and other agricultural products each year.

OVERLEAF
Steaming, sulfurous seepage surrounds an abandoned carbon dioxide plant on the shores of the Salton Sea. Sitting atop the colliding tectonic plates along the San Andreas fault, this area of the Imperial Valley is covered with salt-laden sediments of the Colorado River Delta and is a prime geothermal energy production area.

Created early this century when the Colorado River broke through irrigation canals, causing catastrophic flooding, the Salton Sea is now shrinking. The delta of the New River snakes out into the lake, adding its highly polluted waters to the increasingly salty water. Laden with industrial and medical wastes from Mexico, pesticides and salt from irrigation, the New River is one of the most polluted streams in the United States.

CHAPTER FOUR

MEXICO

Nowhere on the Colorado did I get a greater sense of the dividing of the waters between the haves and have-nots than below Morelos Dam in Mexico, south of Yuma, Arizona. The last of the Colorado River water is pushed into the Canal Central here, and the riverbed becomes shallow enough to wade across. Many Mexicans hoping to start a new life with the water and wealth of the United States have waited in the riverbed until dark before crossing the border.

On my first trip past the Morelos Dam, I spotted several groups of young men crouched on the dry river bank. They found their way up the channel by sighting in on a lighted cross on a Yuma church.

Without a real river, the poor have built makeshift homes along the canal. They are called *avecindados,* squatters. I watched an old man wash his clothes and hang them on a *cachanilla*, or arrowwood, plant. Children played in the canal bottom in the balmy afternoon. A woman named Juana Rosales squatted on the sloping canal side and scrubbed her clothes back and forth on concrete to clean them. She then carried a bucket of water to her garden, a skimpy row of corn and squash. The homes are part adobe, made from canal mud, part cardboard and car parts. The squatters come from the interior of Mexico where economic conditions are worse, hoping for work near the city or a chance to start a new life "over there" in the United States.

For 20 miles the nearly empty riverbed is the border between the United States and Mexico, but metaphorically the Colorado has divided the two countries since the first attempts to divert the river into California. In 1904 Mexico was to receive half the water in the first canal ever built; it ran on both sides of the border. But then the All-American canal was built and Hoover Dam's gate closed. Protesting its loss of water, in 1944 Mexico received the right to 1.5 million acre feet of water, about half of Arizona's or Colorado's share.

When the United States developed the Welton-Mohawk irrigation district east of Yuma, salinity levels in Mexico shot to over 1,000 parts per million, killing crops. Rather than take the saline U.S. soil out of production, the Bureau of Reclamation convinced Congress to authorize a $260 million desalination plant near Yuma. The plant will cleanse the water to the exact salinity level required by the treaty.

Mexico has tried for years to reopen negotiations to get more water. They are running short, and the ecosystem in the delta has been devastated. But under the treaty, Mexican officials don't expect more water from the United States "The states aren't interested in giving up anything," said Luis Lopez Moctezuma, a planning official for Baja California. Only two-thirds of irrigable land in Baja gets water from the river. The water is divided among 14,000 farmers, each of whom can plant only 40 of their 50 acres. "They don't have the water to plant the whole area all year long. Farmers are limited to one crop a year," he said. There also is an increasing demand for urban water: an aqueduct takes Colorado River water across Baja to Tijuana. "Because of the urban growth on the coast and here, people and the farmers are at odds on water use," said Cuauhtemoc Leon Diaz, the head of the regional university museum in Mexicali. "When the treaty was made no one took into account the growth of Baja California."
Mexico will even feel the effects of lining the All-American Canal. The water that now leaks down through the sand is pumped out on the Mexico side and used to irrigate 50,000 hectares. That will dry up after lining. "So the water gained in California is lost to us," he said.

Driving south from Mexicali, the small plots and rusty homestead give way to makeshift homes built along the concrete-lined canals. My host, Charlie Williams, drove through the outback, trying to find the river. The dirt road dipped and Charlie said: "I think that's the river." It was filled with cottonwood and tamarisk. We turned around and drove into the dip again. "I'm pretty sure this is the river." We drove past the foundation of an old school once by the riverbank. Palm trees were planted in a row. Charlie's father took a picture of a flood once, with someone rowing a boat into the school. With the river gone, so went the people. As late as the 1950s there were ferries across the river here. None is needed now.

It was in the river delta that I saw the real effect of the water

In the cotton fields of Ejido Janitzio south of Algodones, young women like Esthela Alamilla wear bandanas to protect their skin from the sun and the desiccated cotton plants.

semillas

The last waters of the Colorado River trickle through the gates of the Morelos Dam on the Mexican border.

The Yuma Desalting Plant will salvage more than 70 million gallons of salt-laden irrigation drainage water a day. The Bureau of Reclamation's $360 million plant uses 9,360 reverse osmosis membrane filters to remove salt from the water.

shortage. The delta was once a series of green lagoons that ecologist Aldo Leopold described as a "milk-and-honey wilderness" where egrets gathered like a "premature snowstorm," jaguars roamed, and wild melons grew. That was in 1922, the year the Colorado River's interstate compact was signed and plans were laid for Hoover Dam. In the years since, Cuauhtemoc Leon Diaz told me, the ecosystem has changed completely – "because of salt, because of less sweet water, because we don't have the sediments which brought nutrients."

Two marine animals became endangered species: a fish called the *totoaba* and a porpoise called the *vaquita*. The *totoaba* used to grow to six feet and weigh 300 pounds; its flesh and bladder were delicacies. Adults migrated to the mouth of the delta in thick schools, and the tides carried the eggs back into the natural nursery of the delta. When the river was cut off, the nursery mechanism was cut off too. The decline began in 1935, the year Hoover Dam closed its gates.

Mexican biologist Oscar Baylon Grecco, chief of the federal fisheries department in Mexicali, said the number of food-chain species in the delta declined from 14 in 1902 to 2 in 1982. As a result the number of fishermen in the delta declined from 300 to 60.

The loss to Mexico is incalculable. Like the burning of a rain forest, a genetic pool of unknown wealth disappeared, with its intertidal vegetation and riparian life. To give an idea of what might have been lost, consider the case of a wheat-like salt grass used by the Cucapa Indians. The last known harvest was 1951 when they gathered it in windrows and threshed it to make bread. The plant was thought to be extinct — until it was rediscovered in the delta by saltwater agronomist Nicholas Yensen. Through selective breeding he improved the yield from one or two pounds an acre to one or two tons an acre. The grass grows best when irrigated with full-strength seawater, making it valuable in arid and saline areas worldwide.

"Here was a grain the Indians used that was thought extinct," said Yensen, who traveled the world on a Rockefeller grant looking at thousands of plants for use in saline conditions. "And the one that made the most sense was the one that was nearly extinct. There is tremendous use for this plant, and others like it. We've developed a forage for animals. We're now working to put the plant in Africa, where people are starving to death with vast, but saline, waters a few feet below the surface."

"When the Colorado stopped flowing, we lost the main population of the plant — it could have gone extinct, and we'd never have known," he said. The river was like the Nile in its importance to the delta. "We probably never will know what we lost."

Mexican biologists have little hope of restoring the ecology of the Colorado delta. "While the Mexican government recognizes the loss of species, there is nothing they can do, given the economic conditions. Even if we got more water from the United States it would be put into agriculture uses, not used to flush the delta. The basic need is for human beings," said Diaz. When asked what message he wanted to convey from the Mexican side, Diaz said: "There should be a consciousness that water treaties are susceptible to change. And that water is a resource for the length of its bed. The United States should share in the life of the delta and the economies of the delta area."

In my last days on the river I talked with Cucapa Indians at their village, aided by Anita Alvarez de Williams, a self-taught expert on the small tribe. She introduced me to several villagers and translated their sad stories. They live on the Rio Hardy, an arm of the Colorado that flows sluggishly, fed by return flows of Mexican irrigation. They cannot drink it; their drinking water comes by tank trucks. No longer do they farm. When the river flooded in the mid 1980s, the government moved them to brick and frame homes on a gravel bar. It had been six months since they had been able to catch enough fish to sell. Still, at every home a boat lay expectantly in the gravel. "I don't have much hope for the future," said Rosendo Carrillo Oliveras "The older ones fish, the younger ones goof off, because there is no work."

Anita introduced me to Inocencia Gonzales Saiz, a 53-year-old woman who makes traditional beaded collars, worn in ceremonies. Instead of fish, she said, the Indians eat frijoles and junk food. The water can't be used for melons or squash because it is too salty. They eat a diet high in sugar and fat. Many of the people have diabetes. A nephew of hers died of alcoholism. "I know three or four who died of alcoholism," she said. "One is my husband. He was a fisherman. He drowned."

Onesimo Gonzales Saiz, the village leader, put it simply: "We've always existed, not poor, just existed. We're still where we where born. Tell the world there are people speaking our language, there are still remnants of our culture. There are still river people."

I thought back to all that I had seen on the Colorado. The river could produce so much — and promise so little. As we left the village of El Mayor, Anita said to me: "There are important things to learn from apparently simple cultures. The fancy cultures and their temples come and go pretty fast. These guys have been around for a couple thousand years. But barring a miracle, you're seeing the last of them."

Aldo Leopold, at the end of his essay on the Colorado Delta, wrote: "Man always kills the thing he loves . . .I am glad I shall never be young without wild country to be young in."

With what water it had, the Colorado created a new civilization in the Southwest. Now it is beset by the needs of mature as well as burgeoning communities. There is talk of renegotiating the 1922 compact to move more water where the people are and of creating new ways to exchange, augment, conserve and manage. Here in the delta, that seemed remote and meaningless. It reminded me of an aging prima donna eking out a little more time, refusing to face the inevitable.

"What river?" El Coyote had asked me.

I knew the answer, but had no reply.

As dusk gathers over Mexico's Canal Central, Juan Garcia and his young friend, Alehandro Lemos, drive their goats home.

Only a few miles south of the Morelos Dam in Mexico the last of the Colorado River sinks into the sands.

Right

Bird tracks and a slowly sinking pool of water in Mexico mark the end of the once mighty Colorado River. Disappearing into the sands and leaving minnows and shrimp flopping in the sun, the river is gone less than five miles south of the U.S.-Mexican border.

Far Right

Mexican irrigation canals bring water to the fields and bath water to Consuelo Garcia and her three-year-old daughter, Dolores Ochoa. Garcia has lived as a squatter for two years on the edge of an onion field. She is like thousands of peasants in the Mexican irrigation district who find life as a landless squatter here better than in the Mexican interior. Bathing in the canals is illegal because of parasites, but it is not possible to enforce such a law.

Great clouds of dust rise from fields where tractors pull laser-guided levelers beside the Canal Central in Mexico. Farmers are limited to about 50 acres of land each, but many circumvent the law to put together much larger holdings. Laser leveling of fields can make flood irrigation much more efficient, assuring that all areas of the field receive equal amounts of water.

ABOVE
Colorado River water grows vast fields of cotton in Mexico, and harvesting still is done by hand. Picking from sunrise until noon (when it gets too hot to work), a strong worker can pick up to 440 pounds of cotton a day, earning $11.

LEFT
A small boy rests in the shade of a cotton plant while his mother picks cotton near Ejido Pachuca. She makes about $6 a day.

FIRST OVERLEAF
An old man struggles to weigh his bag of cotton. A big bag is about 77 pounds, and a good day for a peasant would be six bags. Development of the Colorado River water in Mexico has attracted thousands of the very poor to seasonal work and squatters' houses.

SECOND OVERLEAF
Night falls along the Canal Central in Ejido Pachuca as Trinidad Garcia sits with his children. He lived in Houston for four years and worked for one year before losing his job because he lacked immigration papers. Now he finds life better as a squatter in Mexico. Garcia built his adobe house from bricks he made himself. With enough children, some of whom will inevitably find work in the United States and send money home, his old age will be secure.

10
20
30
40
50
60
70
80
90

BABY

Work comes and goes with the agricultural season for men living along the Canal Central. When there is no work they play cards or enlarge their houses to accommodate growing families.

Evening is bath time along the Canal Central in Ejido Pachuca. Squatters can live on any land along the canal not already taken. Many lived along the Colorado River, a mile to the east, before the Canal Central took all the river's water.

FIRST OVERLEAF
Fingers of the Sea of Cortez reach back into the Colorado River Delta in Mexico. Seawater washes the tidal flats, but it has been eight years since water from the Colorado River reached the sea. At least one species of fish, the totoaba, *depended on the mixture of fresh and tidal waters in its spawning grounds. It is now nearly extinct. Throughout the area a rich wetland habitat has been destroyed.*

SECOND OVERLEAF
Low tide on the Sea of Cortez near El Golfo de Santa Clara brings out residents of the tiny Mexican fishing village to dig for clams in the sediments of the ancient Colorado River Delta.

The soils in the Colorado River Delta, carried here from the Grand Canyon, now dry and crack as the river is stretched beyond its limits.

Following the patterns of his ancestors who have lived in the surrounding mountains for more than 2,000 years, Matias Portillo is one of the last of the Cucapa Indians to fish the Rio Hardy, a nearly fruitless occupation now that the Colorado River no long brings water to freshen its flow. With a face as cracked as a dry riverbed, Portillo lives in the tiny village of El Mayor.

Cucapa women drag their nets out of the Rio Hardy to wait for more water and better fishing.

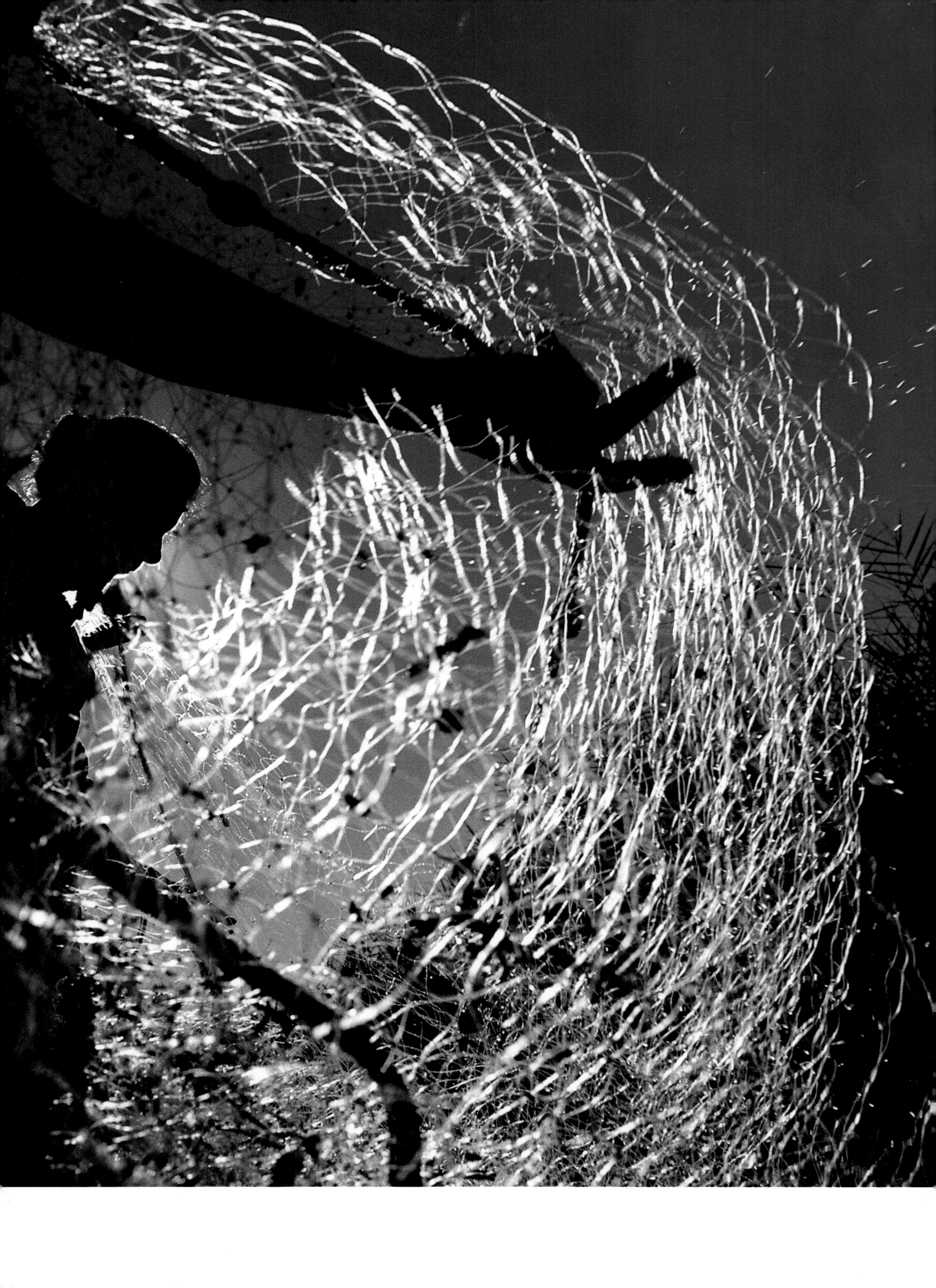

Derelict cars and ruined signs are the playground of Cucapa children.

If he is lucky, fishing all night will bring Arnulfo Golaviz Zais about two pounds of shrimp worth about $14. The Cucapa say the river "...is our life." But here in Mexico there is no more Colorado River.

ACKNOWLEDGMENTS

Countless individuals and organizations assisted us during our research, writing and photography of this book. From the early stirrings of this project at *The Denver Post*, many colleagues have championed our cause. Far too many must go nameless, and be content with our sincere gratitude.

Thanks to all the people at the National Geographic Society for whom the majority of this work was done. Editor Bill Graves first envisioned and approved the project, and Pritt Vesilind saw it through. Picture Editor Dennis Dimick merits special thanks for moral support via long distance.

Among those who assisted us in the field we must mention Bruce Gordon and Project Lighthawk (who gave us our first views of the Colorado River from the air); Chuck Troendle and the U. S. Forest Service; numerous Bureau of Reclamation people, who went out of their way to help us, including Dave Wegener, Gary Kuhstoss, Bob Steele, and Tom Gailey; the scientists and rangers for the U.S. Fish and Wildlife Service, including Harold Tyus and Will Nidecker; the wonderful folks at Grand Canyon Dories; and Martin Litton who showed us the Grand Canyon; and our friends at Navajo Agricultural Products Industry. Others who were generous and kind, include R. Frank Norton, Jim Taussig, Larry Cox and Harvey Johnson (who made one last trip to the Grand Ditch before passing away last year.)

Finally Anita Alverez de Williams of Mexicali went far afield to introduce us to her beloved friends, the Cucupa Indians. Through her kindness we, too, found friends of our own among the Cucupa of El Mayor and among the gracious people who live by the irrigation canals of Ejido Pachuca and the surrounding farming areas of Mexico. They let us into their homes and into their lives.

-Jim Richardson and Jim Carrier

The photographer

TECHNICAL CREDITS

Photographers have a strange, but cozy relationship with the machines they use to create their art. My constant companions during 20,000 miles on the road were my battered Olympus OM-4ts. Though much abused, they never failed me. I fell on one at Lake Powell and cracked a rib, but the camera kept working. Lenses included a16mm Fisheye,18mm, 21mm, 24mm Shift, 28mm, 35mm, 50mm, 90mm Macro, 100mm, 180mm, 350mm and a 70-210mm zoom.

The trunk of my Corolla was crammed with two cases of lights (strobes and tungsten), light stands, my Bogen tripod with an Arca Swiss ball head, clothes, various hats for my balding head, waders, and a cooler full of film. My film choices ranged between Kodachrome 64 and 200, and Fujichrome 50 and Fuji Velvia. Exposure was almost always determined with the wonderful Olympus metering system, and occasionally with a Minolta Flashmeter IV.

Story research was done on my Apple Macintosh SE computer using FileMaker II software. In the course of the story I talked to more than 300 sources. I can't image keeping track of them without my "Mac."

I chartered a variety of aircraft for aerial photography, including Cessna 152, 172, 182 and 210 high wing airplanes, a Robinson helicopter and a Bell Jet Ranger helicopter. I am grateful to all the pilots who's skill put me where I needed to be to get the pictures.

And finally I can't neglect a few things that kept me going. My books on tape (they made the miles fly by), and every Quik Shop, 7-11, Blakes Lot-a-Burger, McDonald's, Carl Jr's, Burger King, and Mr. Burger between Denver and the Sea of Cortez, not to mention the nice guy who sold me hotdogs by the road in Mexico.

-Jim Richardson

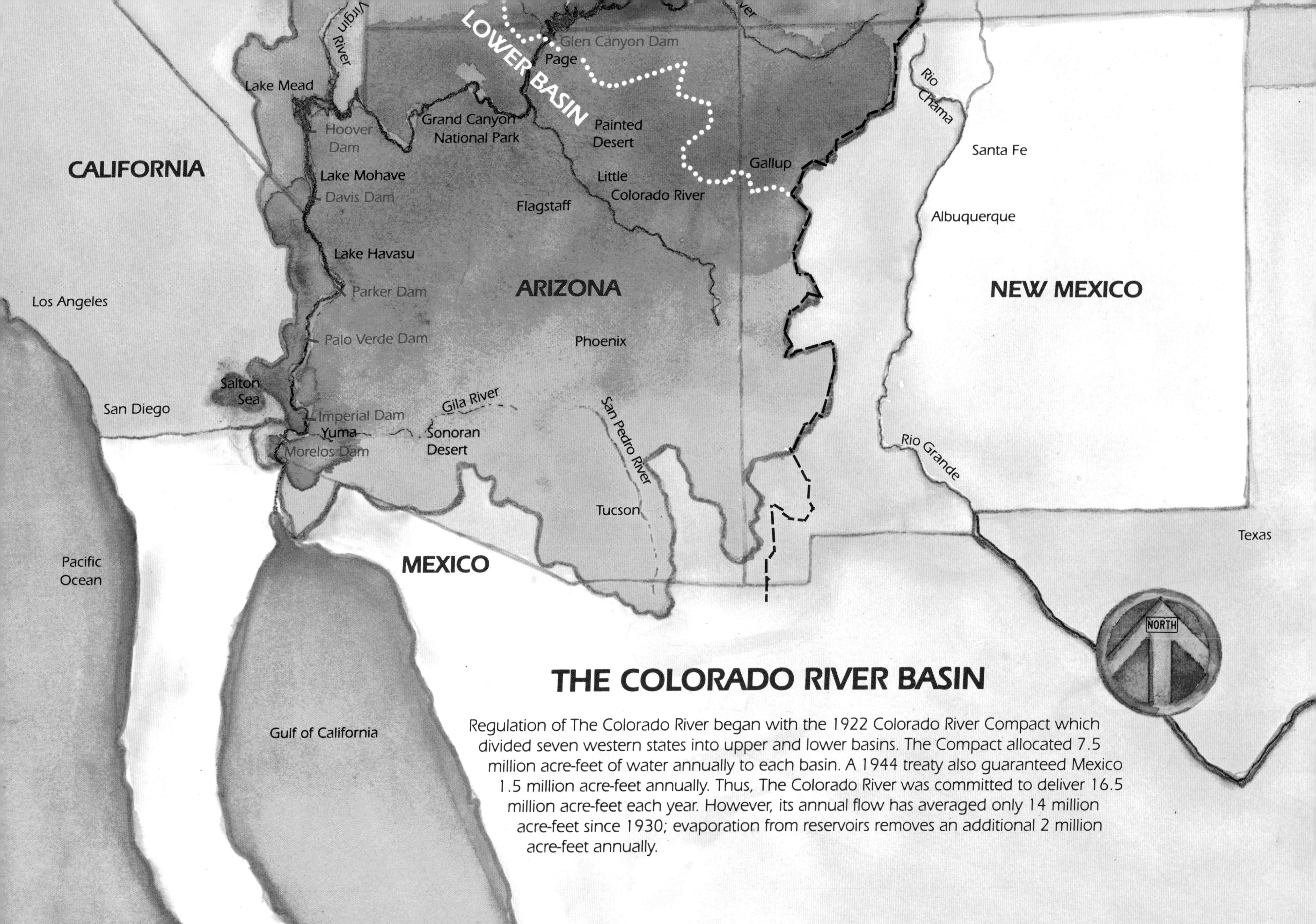

THE COLORADO RIVER BASIN

Regulation of The Colorado River began with the 1922 Colorado River Compact which divided seven western states into upper and lower basins. The Compact allocated 7.5 million acre-feet of water annually to each basin. A 1944 treaty also guaranteed Mexico 1.5 million acre-feet annually. Thus, The Colorado River was committed to deliver 16.5 million acre-feet each year. However, its annual flow has averaged only 14 million acre-feet since 1930; evaporation from reservoirs removes an additional 2 million acre-feet annually.